REVISE EDEXCEL FUNCTIONAL SKILLS LEVEL

Mathematics

REVISION GUIDE

Series Consultant: Harry Smith

Author: Sharon Bolger

A note from the publisher

In order to ensure that this resource offers high-quality support for the associated Pearson qualification, it has been through a review process by the awarding body. This process confirms that this resource fully covers the teaching and learning content of the specification or part of a specification at which it is aimed. It also confirms that it demonstrates an appropriate balance between the development of subject skills, knowledge and understanding, in addition to preparation for assessment.

Endorsement does not cover any guidance on assessment activities or processes (e.g. practice questions or advice on how to answer assessment questions) included in the resource nor does it prescribe any particular approach to the teaching or delivery of a related course.

While the publishers have made every attempt to ensure that advice on the qualification and its assessment is accurate, the official specification and associated assessment guidance materials are the only authoritative source of information and should always be referred to for definitive guidance.

Pearson examiners have not contributed to any sections in this resource relevant to examination papers for which they have responsibility.

Examiners will not use endorsed resources as a source of material for any assessment set by Pearson.

Endorsement of a resource does not mean that the resource is required to achieve this Pearson qualification, nor does it mean that it is the only suitable material available to support the qualification, and any resource lists produced by the awarding body shall include this and other appropriate resources.

Contents

A small bit of small print
Edexcel publishes Sample Test Materials on its website. This is the official content and this book should be used in conjunction with it. The questions in *Now try this* have been written to help you practise every topic in the book. Remember: the real test questions may not look like this.

Online test preparation

If you are taking the online test, you will need to understand how it works. Before you start, read the instructions about how to use the test and make sure you know what all the icons do.

Useful icons

You can click the Time icon to find out how much time you have left on your test. The time will appear in the bottom right-hand corner.

The timer does not stop when you click on the help button. You will be reminded when you have 15 minutes left, and again when you have 5 minutes left in the test.

You can click this Help icon if you want a demonstration of how the online test buttons work.

If you are unsure how to answer a question, click the Flag icon and move on to the next question. Come back to the questions you have flagged later.

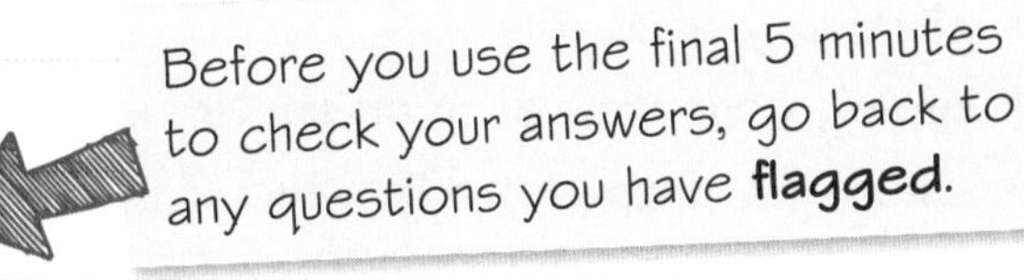

When you click the Review icon, all your flagged questions will appear. To go back and answer one of the questions you have flagged, click on that line.

These buttons move you from question to question.

Be very careful with the Quit button. If you click on it and then click on 'Yes', you will not be able to return to the test even if you haven't finished!

See page 3 for useful tips on how to use the onscreen calculator.

Changing the test settings

Click on the + button in the bottom left-hand corner of the screen to open the settings box.

Use the colour and zoom reset buttons to go back to the original settings.

Click the arrows to move around the page when you are zoomed in.

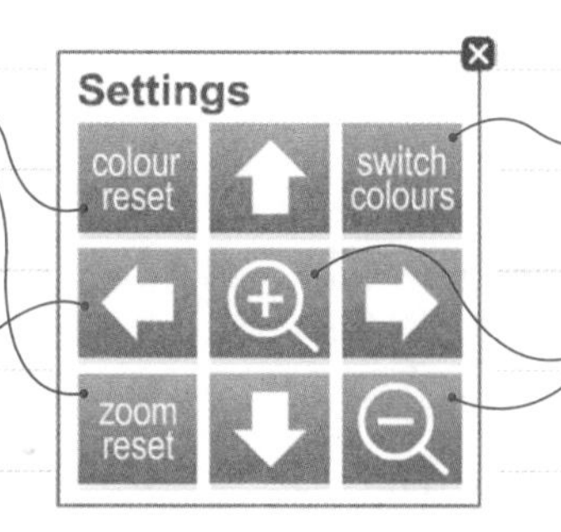

Click the switch colours icon to change the colour of the test to make it easier to read.

Click the magnifying glass icons to zoom in and out.

Now try this

1 How do you find out how much time is left?
2 What can you do if you can't read the test clearly?

Had a look ☐ Nearly there ☐ Nailed it! ☐

Online test tools

If you are taking the online test, it is a good idea to prepare by practising using the online tools to answer questions.

1 Show your working

When asked to show your working, write **every step** of your working out in the working out box. Type the numbers and click on the blue buttons to add in symbols.

When you have worked out the answer, you will need to show this by:

- clicking yes or no
- typing your answer in an answer box
- clicking a drop-down list and selecting the right option
- selecting the right answer from a list of options by putting a tick in the box.

° x^2 √ + - × ÷

You might get marks for working out even if your final answer is wrong.

2 Tables

When you see a table, you could be asked to:

- click a particular row or column
- type information into an empty cell
- drag and drop data into a cell.

To drag and drop:

1 Click on the thing you want to move and drag it to where you want it.

2 Drop it into the empty space.

Playground | Story time | Drawing

	Session 1	Session 2	Session 3
Group A			
Group B			

3 Shape and space

You could be asked to place an object onto a plan. This object represents a home cinema system.

Click and drag the object to the floor plan. Drag the dots at the corners of the object to resize it.

4 Graphs

You could be asked to display information as a graph. Remember to fill in all the missing information.

✓ Type in the graph title.

✓ Type in the axis labels and scales. The answer boxes will expand to fit the text you type in them.

✓ To plot each point, drag the cross to the graph and drop it in the right position. The red line connecting the points will appear automatically.

✓ Check your graph. You can move the points by clicking and dragging them.

Now try this

1 Why is it important to show your working out?

2 How can you prepare for your online test?

Using the onscreen calculator

You are allowed to use a calculator in your test. Make sure you know how to use one.
If you are doing the online test, press the blue button to open the onscreen calculator. 

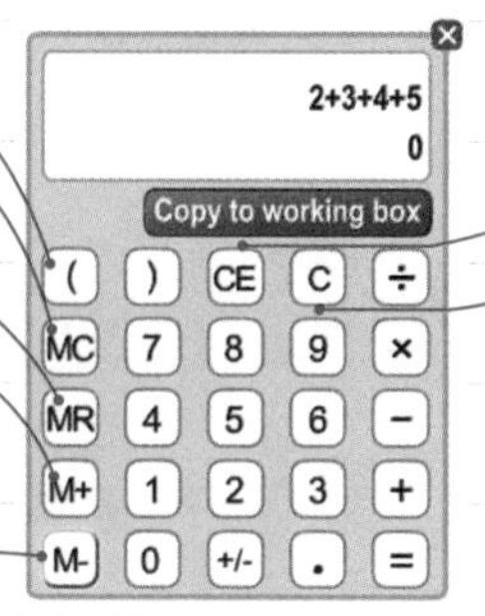

Order of operations

You will sometimes need to do calculations that involve more than one step.

You can enter each step separately or you can type in the whole calculation at once.

If you type in the whole calculation, make sure you use the correct order of operations.

Click 'Copy to working box' to show your working out.

If you are sitting a paper test, you will be allowed to use your own calculator. Make sure you are familiar with how to use it before the test.

Worked example

A nursery teacher buys a toy and a drink for four children. The toy costs 99p and the drink costs 45p.

Work out the total amount the teacher spends.

You need to be careful when using your calculator to do all the steps at once. Make sure you put brackets in the right place and always check your answer.

Method 1 – work out each step separately

Cost for a toy and drink for each child:

99 + 45 = 144p or £1.44

Total cost for four children:

4 × £1.44 = £5.76

Method 2 – use a calculator to do all the steps at once

(99 + 45) × 4 = £5.76

Now try this

For these questions, use your calculator. Write down the buttons you press to do the calculation both in individual steps and all at once.

A carpenter needs to make three tables. The wood for each table costs £26 and the fixings cost £2.

(a) How much will all the tables cost to make?

(b) The carpenter sells each table for £50. How much money does he make?

Had a look ☐ Nearly there ☐ Nailed it! ☐

Number and place value

It is important to know the size of numbers in everyday and work life. A number is made of digits. The more digits a whole number has, the larger the number. The position of the digits is important as it tells you its value and helps you to read the number.

Worked example

1 (a) Write the number 8302 in words.

Eight thousand, three hundred and two

(b) Write the number 121 345 in words.

One hundred and twenty-one thousand, three hundred and forty-five

(c) Write the number fourteen thousand, three hundred and forty-seven in figures.

14 347

(d) Write the number two hundred thousand, three hundred and twenty-six in figures.

200 326

(e) Write the number one million, five hundred thousand and four in figures.

1 500 004

	millions	hundred thousands	ten thousands	thousands	hundreds	tens	units
(a)				8	3	0	2
(b)		1	2	1	3	4	5
(c)			1	4	3	4	7
(d)		2	0	0	3	2	6
(e)	1	5	0	0	0	0	4

The zero means 'no tens'

200 000 or 2 hundred thousands

The zeros fill the empty spaces

Read numbers in groups of three from the right.

1 500 004 reads one million, five hundred thousand and four.

Worked example

2 (a) Write down the value of the digit 3 in:
(i) 2356 **(ii)** 34 610 **(iii)** 230
(iv) 3 521 467

(i) 300 (ii) 30 000
(iii) 30 (iv) 3 000 000

(b) Which of these numbers is larger:
(i) 4096 or 4690? **(ii)** 31 120 or 31 102?

(i) 4690 (ii) 31 120

Compare numbers by looking at the position of the digits. Start from the left side as the value of the digits is larger.

millions	hundred thousands	ten thousands	thousands	hundreds	tens	units
			④	⓪	9	6
			④	⑥	9	0

Start from the largest place value. The digits are the same.

6 hundreds is larger than 0 hundreds. 4690 is larger than 4096

Now try this

1 (a) Write the number 3421 in words.
(b) Write the number three million, five hundred thousand in figures.

2 Write down the value of the 4 in each of these numbers.
(a) 2432 **(b)** 42 321

3 The manager of a company needs to order the amount each member of the sales team earned in one month.
Order these amounts from smallest to largest.
£123,506 £38,003 £42,023 £30,803

Negative numbers

Numbers that are greater than zero are called positive numbers. Numbers that are smaller than zero are called negative numbers.

Number lines

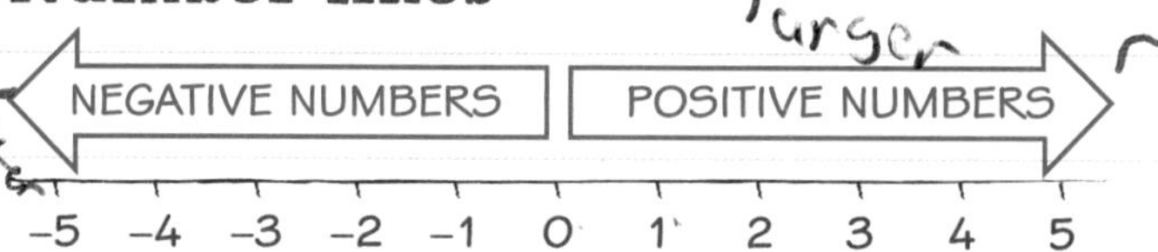

A number line can help you put numbers in order of size.

As you go to the **left** the numbers get **smaller**.

As you go to the **right** the numbers get **bigger**.

0 is not positive or negative.

Identifying negative numbers

A **negative number** has a minus (–) sign before it.

–2, –4, –6 are all negative numbers.

A **positive number** sometimes has a plus (+) sign before it. If a number doesn't have a sign before it, you can assume it is **positive**.

+3, 7, 9 are all positive numbers.

Worked example

A company's cash reserve is how much they have in the bank at the end of each month.
When the balance is negative, the company is overdrawn.

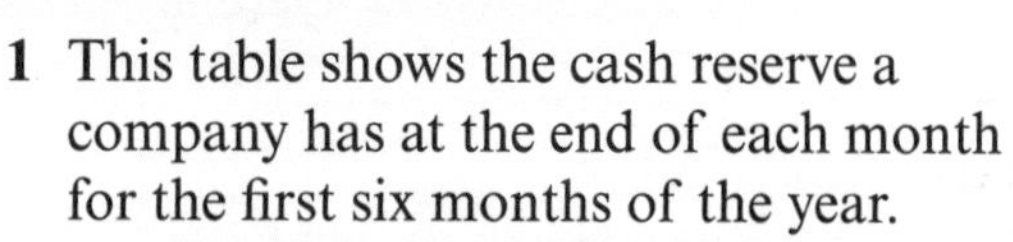

1 This table shows the cash reserve a company has at the end of each month for the first six months of the year.

Month	Balance
January	–£340
February	£810
March	£260
April	–£120
May	–£100
June	–£452

(a) For how many months was the company overdrawn?

four months
January (–£340), April (–£120), May (–£100), June (–£452)

(b) In which month was the company most overdrawn?

June (–£452)

Worked example

2 Below is a list of the average February temperatures of six countries.
–1 °C, –15 °C, –2 °C, 2 °C, 9 °C, –8 °C

(a) Order the temperatures from lowest to highest.

–15 °C, –8 °C, –2 °C, –1 °C, 2 °C, 9 °C

(b) How many countries had an average monthly temperature less than –7 °C?

Two countries (–15 °C and –8 °C)

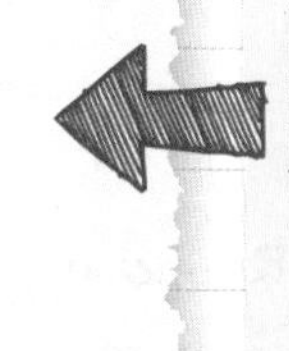

Imagine a number line with all of the values on it. Remember that the further to the left the number is, the lower (colder) it is.

Now try this

1 Write these numbers in order of size. Start with the smallest number.
6, 10, –2, –7, 9, –5

2 In Europe, the temperature of a freezer in a restaurant must be –18 °C or less. Which of these temperatures are less than –18 °C?
–20 °C, –10 °C, 19 °C, 8 °C, –16 °C

Rounding

It is often easier to use rounded numbers in rough calculations.

Rounding to the nearest 10

Look at the multiples of 10 that the number lies between. Round to the closest multiple.
If the number is exactly halfway between the nearest multiples of 10, round up.

46 lies between 40 and 50

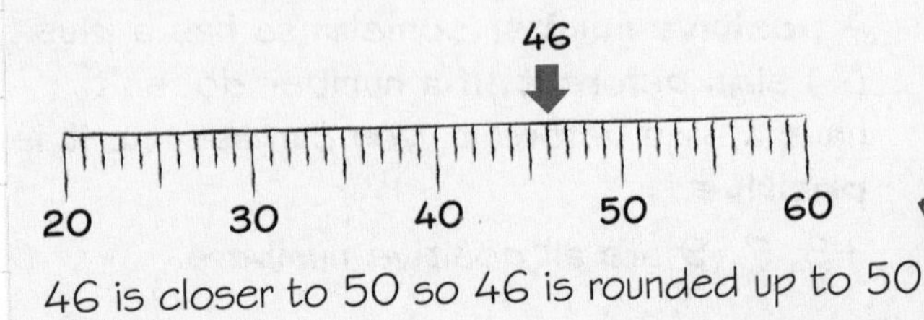

46 is closer to 50 so 46 is rounded up to 50

132 lies between 130 and 140

132 is closer to 130 so 132 is rounded down to 130

Worked example

1 (a) A piece of wood measures 46 cm. Round this to the nearest 10 cm.

46 cm rounded to the nearest 10 cm is 50 cm.

(b) A piece of wood measures 132 cm. Round this to the nearest 10 cm.

132 cm rounded to the nearest 10 cm is 130 cm.

Rounding to the nearest 100 or 1000

Rounding to 100 or 1000 works the same as rounding to 10

Look at the nearest multiples of 100 or 1000 and decide which the number is closer to.

Golden rule

	Round up	Round down
Nearest 10	5 or more	less than 5
Nearest 100	50 or more	less than 50
Nearest 1000	500 or more	less than 500

Worked example

2 Last month, 4241 people attended a show at a theatre.
Round this number to the nearest 1000

4241 rounded to the nearest 1000 is 4000

4241 lies between 4000 and 5000
4241 is closer to 4000 so 4241 is rounded down to 4000

Now try this

1 Round these numbers to the nearest 100:
(a) 147 **(b)** 365 **(c)** 9123

2 A shop made a profit of £38,827 last year. Round this number to the nearest 1000

Adding and subtracting

For some questions, you need to decide which calculation to do.

Worked example

1 A waiter has two customers.
The first customer leaves a tip of £8. The second customer leaves a tip of £15.

(a) How much does the waiter receive in tips altogether?

8 + 15 = £23

(b) The waiter uses his tips to buy a drink which costs £2. How much money does he have left?

23 − 2 = £21

You need to decide whether to add or subtract.
Use your calculator to work out the answer.

You can check your answer using inverse calculations.

(a) 8 + 15 = 23 23 − 8 = 15 ✓

(b) 23 − 2 = 21 21 + 2 = 23 ✓

Worked example

2 Martha is paid £2,000 per month after tax.
Her bills are:

Rent	£720
Electricity and gas	£82
Council tax	£95
Telephone	£32
Water	£28

How much money will she have left when she pays her bills?

720 + 82 + 95 + 32 + 28 = 957
£2,000 − £957 = £1,043
Martha will have £1,043 left.

Problem solved!

✓ To work out how much Martha spends on bills, you need to **add** all the amounts.
✓ To work out how much she has left, you need to **subtract** the bill total from the amount Martha is paid.
✓ You should also check your answer by seeing if it is sensible.

Read more about checking your answers on page 12.

Now try this

1 Albert sells a motorbike for £2,349 and a helmet for £94. In total, how much money does Albert make?

2 In one day, a cafe makes £923. Its outgoings are £437. How much profit does the cafe make that day?

Outgoings are the amount that is spent on goods or staff.

3 The table shows the number of clients a plumbing business had each week in February.

Week 1	123
Week 2	75
Week 3	131
Week 4	112

What is the total number of clients the plumbing business had in February?

To find the total add up all the numbers.

Had a look ☐ Nearly there ☐ Nailed it! ☐

Multiplying and dividing by 10, 100 and 1000

You will need to know how to multiply and divide whole numbers by 10, 100 and 1000

Multiplying by 10, 100, 1000

	ten thousands	thousands	hundreds	tens	units
				3	2
× 10			3	2	0
× 100		3	2	0	0
× 1000	3	2	0	0	0

Fill the empty spaces with zeros.

When multiplying by 10, each digit moves one place to the left.

When multiplying by 100, each digit moves two places to the left.

When multiplying by 1000, each digit moves three places to the left.

Worked example

A magnifying glass makes items appear 10 times as big as they really are. An insect is 4mm long. How long does it look in the magnifying glass?

$4 \times 10 = 40$ mm

10 times bigger than 4 mm means 10×4 mm

Dividing by 10, 100, 1000

	thousands	hundreds	tens	units		tenths
	2	4	0	0		
÷ 10		2	4	0		
÷ 100			2	4		
÷ 1000				2	.	4

When dividing by 10, each digit moves one place to the right.

When dividing by 100, each digit moves two places to the right.

When dividing by 1000, each digit moves three places to the right.

Now try this

1 Work out: **(a)** 48×10 **(b)** 360×1000 **(c)** $520 \div 100$

2 A garage owner needs to order some stock. The table shows the cost of the items she needs.
She orders 100 of each item. Calculate the total bill.

Item	Cost
engine oil	£35
car mats	£15
timing chains	£21

Work out the total cost for each item and then add them up.

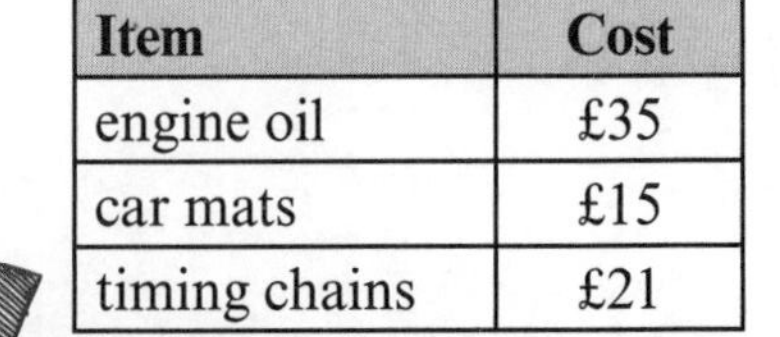

Multiplication and division

You need to be able to decide whether to multiply or divide to answer a problem.

Worked example

1 (a) There are 25 chocolate bars in a box and a shop buys 8 boxes.
How many chocolate bars does the shop buy?

$25 \times 8 = 200$ chocolate bars

(b) Sarah makes 30 cakes. She packs them in boxes that hold 6 cakes each. How many boxes does she need?

$30 \div 6 = 5$ boxes

You can check your answer using the inverse calculation.
$25 \times 8 = 200$ so $200 \div 8 = 25$ ✓
$30 \div 6 = 5$ so $5 \times 6 = 30$ ✓

Worked example

2 The cost price of a USB stick is £7. A shop buys 36 USB sticks and sells them at £12 each.
Work out how much profit the shop makes.

Profit for each stick = £12 − £7
= £5

Total profit = 36 × £5
= £180

Problem solved!

✓ Work out the profit made for one stick.
profit = sales price – cost price
✓ Multiply the profit for one stick by the number of sticks sold.
✓ Don't forget to write your answer in pounds (£).

Worked example

3 Freddie works 40 hours per week and gets paid £320.

(a) How much does he get paid per hour?

£320 ÷ 40 = £8

(b) For overtime, Freddie gets paid twice his normal hourly rate. Last month Freddie did 6 hours overtime. What was his overtime pay?

£8 × 2 = £16

£16 × 6 = £96

Problem solved!

(a) The question asks about how much he gets paid per hour, so you need to divide.
(b) Double means 'multiply by 2'. Work out his overtime wage and multiply it by the number of overtime hours he does.

Now try this

First work out the profit for one camera.

1 252 chairs are to be set up in seven equal rows. How many chairs need to be in each row?
2 A coach can seat 52 people. How many people can five coaches seat?
3 A shop buys cameras for £18 and sells them for £40. The shop sold six cameras in one day.
How much profit did it make that day?

Squares and multiples

You need to be able to recognise and use square numbers and multiples.

Square numbers

When a number is multiplied by itself, the answer is a square number. You can write **square numbers** using index notation.

Multiplication	Index notation	Square number
2 × 2	2^2	4
5 × 5	5^2	25
9 × 9	9^2	81
13 × 13	13^2	169

Try multiplying different numbers by themselves. If the answer is less than 40, try a higher number. If the answer is greater than 60, try a lower number.

Worked example

1 Which of these numbers are square numbers?
1, 36, 53, 100, 92

1 × 1 = 1 so 1 is a square number.

6 × 6 = 36 so 36 is a square number.

10 × 10 = 100 so 100 is a square number.

2 Which square number is larger than 40 but smaller than 60?

49 (7 × 7 = 49)

Multiples

The multiples of a number are all of the numbers in its times table.

A common multiple is a number that is a multiple of two or more numbers.

The multiples of 4 are 4, 8, (12), 16, 20, (24)..

The multiples of 6 are 6, (12), 18, (24)....

12 and 24 are common multiples of 6 and 4 because they are in both lists.

12 is called the lowest common multiple as it is the smallest number in both lists.

Worked example

3 To celebrate its anniversary, a hardware shop gives every sixth customer a free light bulb, and every eighth customer a free pack of batteries.
Which customer will be the first to get both free gifts?

Customers who receive a light bulb

6, 12, 18, (24), 30, 36

Customers who receive a pack of batteries

8, 16, (24), 32

The 24th customer will receive both a light bulb and a pack of batteries.

Problem solved!

Think about how you are going to use multiples to solve this problem.

- ✓ List the multiples of 6
- ✓ List the multiples of 8
- ✓ Identify the first number that is on both lists. This is the lowest common multiple of 6 and 8

Now try this

1 Which of these numbers are square numbers? 4, 12, 16, 50, 54

2 A house alarm sounds every 5 seconds, and a car alarm sounds every 6 seconds. If they initially sound at the same time, after how many seconds will they next both sound at the same time?

Estimating

Estimating the answer to a calculation is useful when you don't need to know the exact answer. Estimating will give you an answer a little bit more or a little bit less than the exact answer.

Estimation

When you estimate, you need to round the numbers first.

Rounding the numbers makes the calculations easier to do.

Round the numbers to the nearest 10 or 100, and then do the calculation.

Symbol for estimation

The symbol ≈ means 'approximately equal to'.

487 ≈ 500 means 487 is **approximately equal** to 500

Worked example

1 (a) Work out an estimate for 58 + 31

Round 58 and 31 to the nearest 10

58 ≈ 60

31 ≈ 30

58 + 31 ≈ 60 + 30

≈ 90

(b) Work out an estimate for 429 − 287

Round 429 and 287 to the nearest 100

429 ≈ 400

287 ≈ 300

429 − 287 ≈ 400 − 300

≈ 100

If the number is between 100 and 999, round to the nearest 100

Worked example

2 A shop sells T-shirts at £36 each and scarves at £21 each.

John buys 12 T-shirts and 12 scarves. Estimate how much this will cost.

Cost = 12 × (36 + 21)

≈ 10 × (40 + 20)

≈ 10 × 60

≈ 600

The cost is approximately £600

Problem solved!

Show the examiner that you have rounded each value.

12 ≈ 10

36 ≈ 40

21 ≈ 20

Use these values to find an estimate of the cost.

Now try this

First estimate how much profit he made on selling one pair of scissors.

1 Estimate the answer to:

(a) 58 + 71 **(b)** 892 – 312 **(c)** 48 × 9

2 Peter sold 89 pairs of scissors for £24 each. They cost him £11. Estimate how much profit he made.

Checking your answer

It is easy to press the wrong button on your calculator and make a mistake. You should always check your answer is correct.

✓ In the test, some questions will have this check symbol next to them. This means that you will be awarded marks for showing how you have checked your answer.

Using estimation

You can use estimation before or after doing a calculation to check if your answer is sensible.

See page 11 for more on estimating answers.

Make sure you show what you have rounded each number to.

Worked example

1 Ashia thinks the answer to 599 + 721 − 421 is 1741
Use estimation to work out if she is correct.

$599 \approx 600$, $721 \approx 700$, $421 \approx 400$

$600 + 700 - 400 = 900$

This is a lot lower than 1741 so she has made a mistake in her calculation.

Using inverse operations

You can check if your answer is correct by using inverse operations.

- Adding and subtracting are inverse operations.
- Multiplying and dividing are inverse operations.

Worked example

2 Dilek thinks the answer to 400 − 221 is 179
Is he correct?

$400 - 221 = 179$

so $179 + 221 = 400$

Dilek is correct.

You should always get the number you started with.

Worked example

3 Cecily thinks the answer to 43 × 5 is 228
Is she correct?

If $43 \times 5 = 228$ then $228 \div 5$ must equal 43

$228 \div 5 = 45.6$

Cecily is not correct.

You could also check by working out $228 \div 43$
You will have a calculator in the test so you can use it to check your answer.

Now try this

1 Dmitri works out 823 + 283 – 789 and gets the answer 1895
Estimate the answer to the calculation. Is Dmitri correct?

2 Use inverse operations to check that these are correct. For any that are incorrect, write down the correct answers.

(a) $423 + 239 = 662$ **(b)** $39 + 323 = 335$ **(c)** $53 \times 31 = 1590$ **(d)** $872 \div 2 = 436$

Word problems

When you are solving problems, you need to:

- ✓ read the question
- ✓ decide which calculation you are going to use
- ✓ check your answers
- ✓ make sure you have answered the question asked.

Worked example

1 The table shows the cost of entrance to a theme park.

adults	£25
children	£12
family (2 adults and 2 children)	£60

Mr and Mrs Svaza want to take their three children to the theme park. Work out the cheapest price they could pay.

25 + 25 + 12 + 12 + 12 = £86

60 + 12 = £72

Problem solved!

The question asks for the smallest amount. The total cost of two adult tickets and two child tickets is £86

The total cost of one family ticket and one child ticket is £72, which is lower. £72 is the cheapest price they can pay.

Worked example

2 Joseph is looking for a job.
He sees these two adverts online.

Advert A	Advert B
telephone salesperson £280 per week 40 hours per week	telephone salesperson £8 per hour 35 hours per week

(a) Compare the pay for the two adverts.

Advert A
pay per hour: £7
weekly pay: £280

Advert B
pay per hour: £8
weekly pay: £280

(b) Which job is better paid?

Advert B is better paid because there are fewer working hours but the same weekly pay.

Problem solved!

(a) Work out the pay per hour. Divide the weekly wage by the number of hours worked.
Work out the pay per week. Multiply the pay per hour by the number of hours worked.

(b) To decide which job is better paid, you need to look at the pay per week and the pay per hour. As the pay per week is the same, compare the pay per hour.

Now try this

Matthew is planning a trip to the zoo for 5 adults and 10 children. The table shows the entrance costs and travel costs. There is a budget of £400. Is this enough?

Zoo		Train fare	
adult	£18	adult	£16
child	£12	child	£8

Fractions

A fraction is part of a whole. A fraction is made of equal parts.

Here is the flag of Ireland:

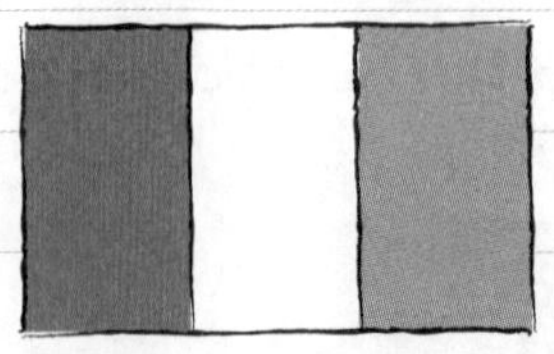

1 part is green → 1 ← numerator
out of → —
3 equal parts → 3 ← denominator

Writing fractions

A pizza is cut into six **equal** pieces.

One piece of pizza = $\frac{1}{6}$

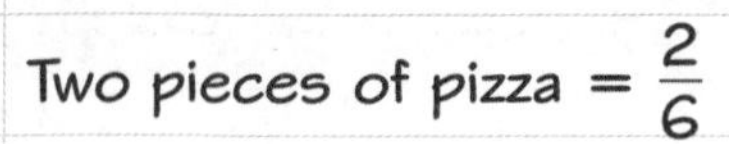

Two pieces of pizza = $\frac{2}{6}$

The whole pizza = $\frac{6}{6}$ or 1

Recognising fractions in words

You need to be able to recognise fractions written in words.

a half = $\frac{1}{2}$ a third = $\frac{1}{3}$

a quarter = $\frac{1}{4}$ a tenth = $\frac{1}{10}$

Worked example

(a) What fraction of this shape is shaded?

$\frac{1}{5}$ is shaded

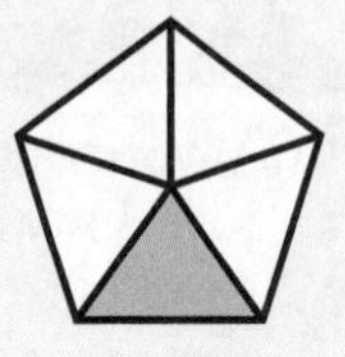

(b) What fraction of the shape is unshaded?

$\frac{4}{5}$ is unshaded

(a) The diagram is divided into **five** equal parts so the denominator is **5**. **One** part is shaded so the numerator is **1**

(b) The number of parts that are not shaded is **four**, so the numerator is **4**

The shaded parts plus the unshaded parts make up the whole: $\frac{1}{5} + \frac{4}{5} = \frac{5}{5}$

Now try this

1 What fraction of this shape is:

(a) shaded black $\frac{2}{5}$

(b) shaded grey $\frac{1}{5}$

(c) not shaded? $\frac{2}{5}$

2 Write $\frac{2}{3}$ in words.

Two thirds

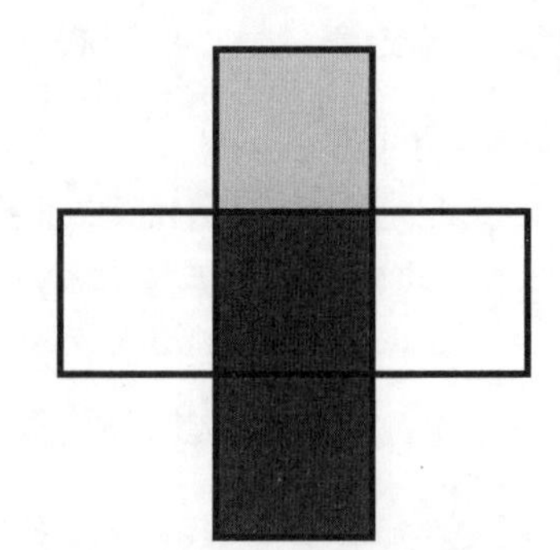

Equivalent fractions

You can compare fractions by using diagrams or equivalent fractions.

Equivalent fractions

Different fractions can describe the same amount. These are called equivalent fractions.

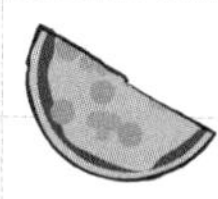

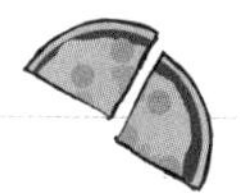

$\frac{1}{2}$ of a pizza is the same amount as $\frac{2}{4}$ of a pizza.

You can find equivalent fractions by multiplying or dividing the numerator and denominator by the same number.

$$\frac{2}{5} \underset{\times 2}{\overset{\times 2}{=}} \frac{4}{10} \underset{\div 3}{\overset{\div 3}{=}} \frac{6}{15}$$

Fraction wall

A fraction wall is a good way of seeing equivalent fractions.

1					
$\frac{1}{2}$	$\frac{1}{2}$				
$\frac{1}{3}$	$\frac{1}{3}$	$\frac{1}{3}$			
$\frac{1}{4}$	$\frac{1}{4}$	$\frac{1}{4}$	$\frac{1}{4}$		
$\frac{1}{5}$	$\frac{1}{5}$	$\frac{1}{5}$	$\frac{1}{5}$	$\frac{1}{5}$	
$\frac{1}{6}$	$\frac{1}{6}$	$\frac{1}{6}$	$\frac{1}{6}$	$\frac{1}{6}$	$\frac{1}{6}$

Simplifying fractions

If you are asked to **simplify** a fraction, divide the numerator and the denominator by the same amount.

If there isn't a number that divides into both the numerator and the denominator, the fraction is in its **simplest form.**

$$\frac{18}{27} \underset{\div 9}{\overset{\div 9}{=}} \frac{2}{3}$$

Comparing fractions

You can compare fractions by writing them as equivalent fractions with the same denominator. Then compare the numerators to decide which is larger.

Worked example

A painter had some paint left in 1 litre tins. The fractions of paint he had left in the tins were: $\frac{1}{3}, \frac{2}{9}, \frac{5}{9}$. Order the fractions from smallest to largest.

$$\frac{1}{3} \underset{\times 3}{\overset{\times 3}{=}} \frac{3}{9}$$

$\frac{2}{9}, \frac{1}{3}, \frac{5}{9}$

Change $\frac{1}{3}$ into ninths so you can compare it to the other fractions.

Now try this

1 Copy and complete these equivalent fractions:

(a) $\frac{1}{7} = \frac{\square}{14}$ **(b)** $\frac{1}{2} = \frac{3}{\square}$ **(c)** $\frac{3}{8} = \frac{\square}{24}$

2 Using the numbers 1, 4 and 8 what is the smallest fraction you can make?

3 Order these fractions from smallest to largest: $\frac{3}{10}, \frac{4}{10}, \frac{1}{5}$

Mixed numbers

Mixed numbers have a whole number part and a fraction part.

$3\frac{1}{4}$ is the same as $3 + \frac{1}{4}$

Improper fractions have a numerator larger than their denominator.

$\frac{5}{2}$ and $\frac{21}{5}$ are both improper fractions.

It is useful to be able to convert between mixed numbers and improper fractions.

Converting a mixed number to an improper fraction

To convert a mixed number to an improper fraction...

Multiply the whole number...

...by the denominator...

...and add it to the numerator.

$$3\frac{1}{4} = \frac{(3 \times 4) + 1}{4} = \frac{13}{4}$$

Keep the same denominator.

Worked example

1 Change each mixed number into an improper fraction.

(a) $4\frac{5}{6}$ $\frac{(4 \times 6) + 5}{6} = \frac{29}{6}$

(b) $3\frac{4}{9}$ $\frac{(3 \times 9) + 4}{9} = \frac{31}{9}$

Multiply the whole number by the denominator, add the numerator.

Converting an improper fraction to a mixed number

To convert an improper fraction to a mixed number...

Divide the numerator...

...by the denominator.

$$\frac{23}{5} = 23 \div 5 = 4\frac{3}{5}$$

Write the remainder as the numerator.

Keep the same denominator.

Worked example

2 Change each improper fraction into a mixed number.

(a) $\frac{34}{5}$ $34 \div 5 = 6 \text{ r } 4$ so $\frac{34}{5} = 6\frac{4}{5}$

(b) $\frac{32}{9}$ $32 \div 9 = 3 \text{ r } 5$ so $\frac{32}{9} = 3\frac{5}{9}$

Divide the numerator by the denominator.

Now try this

1 Work out $\frac{2}{3} + \frac{2}{3}$. Write your answer as an improper fraction.

2 Write $5\frac{5}{9}$ as an improper fraction.

3 Write $\frac{45}{6}$ as a mixed number.

Fractions of amounts

You need to be able to find fractions of amounts.

Fractions of amounts

If the numerator is 1, divide by the denominator.

£40	£40	£40

← £120 →

$\frac{1}{3}$ of £120

£120 ÷ 3 = £40 — Divide by the denominator.

If the numerator is greater than 1, divide by the denominator and multiply by the numerator.

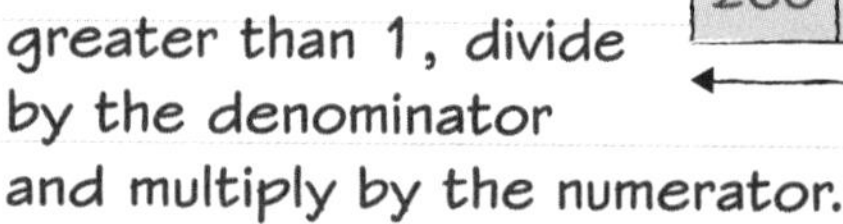

$\frac{3}{4}$ of £240

£240 ÷ 4 = £60 — Divide by the denominator.

£60 × 3 = £180 — Multiply by the numerator.

Worked example

1 Work out: **(a)** $\frac{1}{4}$ of £80 **(b)** $\frac{2}{5}$ of 20 cm

(a) £80 ÷ 4 = £20

(b) 20 cm ÷ 5 = 4 cm
4 cm × 2 = 8 cm

Divide by 5 to find $\frac{1}{5}$, then multiply by 2 to find $\frac{2}{5}$

Worked example

2 Simon wants to buy a table that he has seen on sale in two different shops.

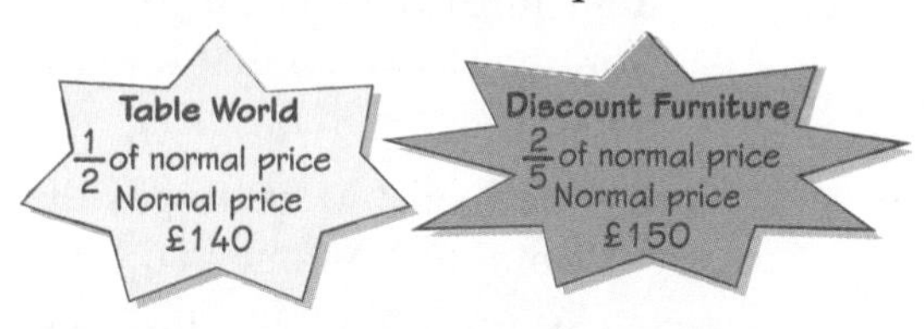

He wants to pay the least amount of money. From which shop should he buy the table? Give your reasons.

Table World
$\frac{1}{2}$ of £140
£140 ÷ 2 = £70

Discount Furniture
$\frac{2}{5}$ of £150
£150 ÷ 5 = £30
£30 × 2 = £60

Simon should buy his table at Discount Furniture as it has the cheapest price.

Problem solved!

Remember to show all of your working clearly so that the reasons for your decision are clear.

- ✓ Work out how much the table costs at each store.
- ✓ Show your working out for each store clearly.
- ✓ Check your calculations to make sure you haven't made a mistake.
- ✓ Make a decision based on your calculations. Write a sentence to explain your reasons.

Now try this

1 Work out:
(a) $\frac{1}{5}$ of £70 **(b)** $\frac{2}{5}$ of £70 **(c)** $\frac{2}{3}$ of 48 g

Use your answer to part (a) to help with part (b).

2 Which is smaller, $\frac{4}{9}$ of £72 or $\frac{3}{4}$ of £44?

 Had a look ☐ Nearly there ☐ Nailed it! ☐

Word problems with fractions

When you are solving problems, you need to:

- read the question
- check your answers
- decide which calculation you are going to use
- make sure you have answered the question asked.

Worked example

1 A theatre had 240 tickets for a show.
On Saturday, it sold $\frac{2}{3}$ of the tickets.
On Sunday, it sold $\frac{1}{4}$ of the tickets.
How many tickets were sold altogether?

Saturday
$\frac{2}{3} \times 240 = 240 \div 3 \times 2$
$= 160$

Sunday
$\frac{1}{4} \times 240 = 240 \div 4$
$= 60$

Total ticket sales
$160 + 60 = 220$

Problem solved!

Plan your strategy before you start.
Break the problem down into steps and lay your work out clearly.

- ✓ Work out the ticket sales for Saturday.
- ✓ Work out the ticket sales for Sunday.
- ✓ Work out the total sales for Saturday and Sunday.

Worked example

2 Nathaniel wants to buy some jeans.
He sees the same pair advertised in two different shops.

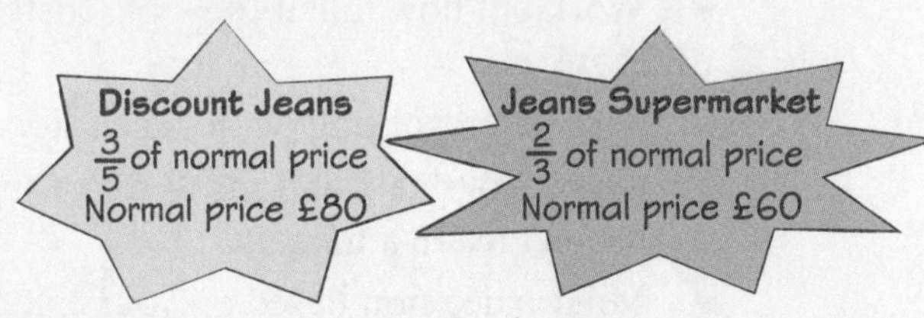

He wants to pay the least amount of money.
From which shop should he buy the jeans? Give your reasons.

Discount Jeans
$\frac{3}{5}$ of £80
£80 ÷ 5 = £16
£16 × 3 = £48

Jeans Supermarket
$\frac{2}{3}$ of £60 = £60 ÷ 3 = £20
£20 × 2 = £40

Nathaniel should buy his jeans at Jeans Supermarket as it has the cheapest price.

Problem solved!

Remember to show all your working clearly so that the reasons for your decision are clear.

- ✓ Nathaniel wants to pay the least amount of money, so you need to work out how much the jeans cost at each shop.
- ✓ Show your working out for each shop clearly.
- ✓ Check your calculations to make sure you haven't made a mistake.
- ✓ Make a decision based on your calculations. Write a sentence to explain your reasons.

Now try this

David buys 80 boxes of chocolates for £200. He sells $\frac{3}{4}$ of the chocolates for £4 each. He then sells the remaining chocolates for £2 each. Work out the total profit that David makes.

Decimals

You will need to be able to order decimals in lots of different types of questions including ordering money, ordering lengths and ordering weights.

Place value diagrams

You can use a place value diagram to help you understand and compare decimal numbers. Remember that decimal numbers with more digits are not necessarily bigger. Try writing extra zeros so that all the numbers have the same number of decimal places.

units		tenths	hundredths	thousandths
0	.	7	5	8
0	.	7	6	0
0	.	7	9	0
0	.	8	0	0

The value of 5 in this number is 5 hundredths.

0.76 is the same as 0.760

0.760 is bigger than 0.758 because 6 hundredths is bigger than 5 hundredths.

0.79 is smaller than 0.8 because the digit in the tenths place is smaller.

Worked example

1 Write down the value of the digit 4 in:

(a) 2.46 — 4 tenths

(b) 32.843 — 4 hundredths

(c) 9.354 — 4 thousandths

2 Which of these numbers is larger?

(a) 0.042 or 0.402 — 0.402

(b) 1.003 or 1.03 — 1.03

3 A carpenter has some lengths of wood. Order the lengths from smallest to largest.

0.63 m, 0.063 m, 0.306 m, 0.36 m, 0.036 m

0.036 m, 0.063 m, 0.306 m, 0.36 m, 0.63 m

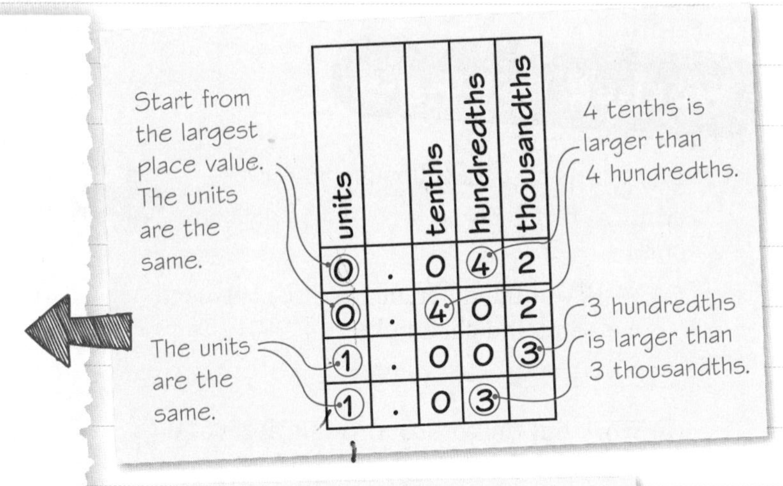

Compare the digits in each place value position.

Now try this

1 For each pair of numbers, decide which number is larger.

(a) 6.7 and 6.456 **(b)** 23.819 and 23.84 **(c)** 2.03 and 2.003

2 What is the value of the 6 in these numbers?

(a) 42.036 **(b)** 2.603

3 A beauty wholesalers sells shampoo by the litre. Here are the prices for four brands of shampoo. Order the amounts from smallest to largest.

£12.30 £12.03 £13.02 £10.30

Had a look ☐ Nearly there ☐ Nailed it! ☐

Decimal calculations

You need to be able to add and subtract decimals up to 2 decimal places. You may need to convert money from pence to pounds before using your calculator.

Worked example

1 Here is a price list for a cafe.

Cup of coffee	£1.10
Cup of tea	90p
Sandwich	£1.99
Biscuit	98p

(a) Find the cost of a cup of tea and a sandwich.

0.90 + 1.99 = £2.89

(b) Find the cost of a cup of coffee and a biscuit.

1.10 + 0.98 = £2.08

The cost of a cup of tea is in pence, and the cost of a sandwich is in pounds.
You need to make sure the costs are in the same units before doing any calculations.

Method 1

The cost of a cup of tea is 90p which is the same as £0.90

0.90 + 1.99 = £2.89

Method 2

The cost of a sandwich is £1.99, which is the same as 199p.

90 + 199 = 289p

289p = £2.89

Worked example

2 Bus tickets cost the following amounts:

adult £1.52
child 65p

(a) Work out the difference in price between the adult and child ticket.

£1.52 − £0.65 = £0.87

(b) Work out the cost of an adult ticket and two child tickets.

£1.52 + £0.65 + £0.65 = £2.82

Convert the child ticket price from pence into pounds so that both prices are in the same units.
To convert from pence to pounds, divide by 100.

Now try this

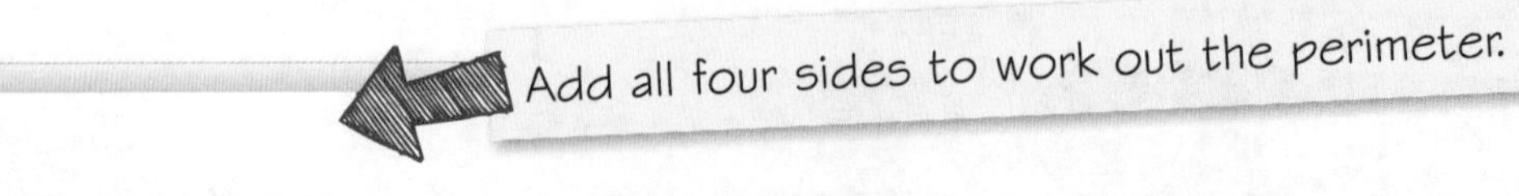

1 Find the difference between £4.92 and 53p.

2 What is the perimeter of this rectangle?

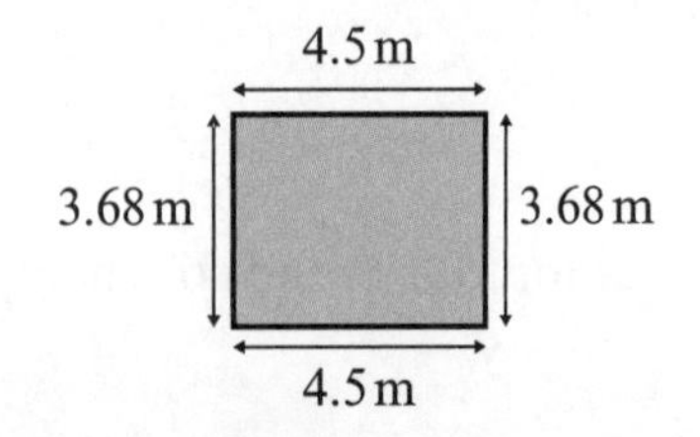

3 Oliver and Lukas go to the shops. They both write a list of how much they spend in each shop. Who spent more money in total?

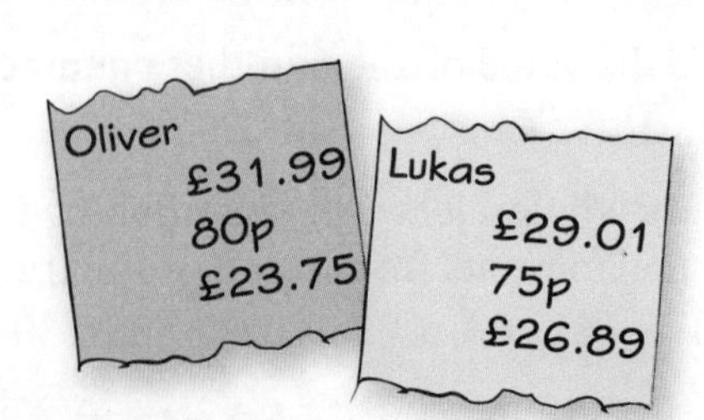

Rounding decimals and estimating

Rounding decimals to whole numbers is useful when estimating the answer to a calculation. When you go shopping, you can estimate the total cost of your shopping bill by rounding the cost of each item to the nearest pound.

2.6 lies between 2 and 3

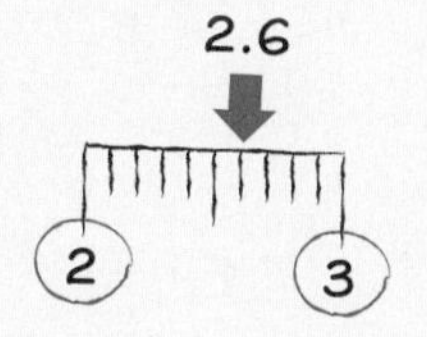

2.6 is closer to 3, so 2.6 rounds up to 3

When the number is exactly halfway between the whole numbers, always round up.

Worked example

1 **(a)** The length of a car is 2.6 m. Round this to the nearest metre.

2.6 rounded to the nearest whole number is 3 so the answer is 3 m

(b) In a shop a coat cost £68.50. Round this to the nearest pound.

68.50 rounded to the nearest whole number is 69 so the answer is £69

Using ≈

The ≈ symbol means **'approximately equal to'**. Use it when rounding and estimating.

Round 7.92 and 2.43 to the nearest whole number.

Worked example

2 Work out an estimate for 7.92 – 2.43

$7.92 \approx 8$

$2.43 \approx 2$

$7.92 - 2.43 \approx 8 - 2 \approx 6$

Worked example

3 Here is Jacqui's shopping bill.

Estimate the total cost of her shopping.

£2.92
£5.31
£4.83
£2.42

3 + 5 + 5 + 2 = £15

Problem solved!

Show that you have rounded each value.

$2.92 \approx 3$

$5.31 \approx 5$

$4.83 \approx 5$

$2.42 \approx 2$

Use these values to find an estimate of the cost.

Now try this

1 Round these numbers to the nearest whole number.

(a) 4.2 **(b)** 15.74 **(c)** 9.8

2 Estimate the answer to:

(a) 3.7 + 8.2 **(b)** 2.85 + 7.92 – 4.3

Had a look ☐ Nearly there ☐ Nailed it! ☐

Word problems with decimals

There are some methods you can use to help you solve different types of word problems.

Worked example

1 Simon is working out how much it will cost to decorate a room.

He needs 12 rolls of wallpaper and 5 litres of paste.

A roll of wallpaper costs £23.40

A litre of paste costs £3.24

He has a budget of £300. Does he have enough money to decorate the room?

12 × £23.40 = £280.80

5 × £3.24 = £16.20

£280.80 + £16.20 = £297

Yes, he does have enough to wallpaper the room.

Problem solved!

If you come across a tricky or unfamiliar question in your exam, you can try some of these methods:

- ✓ Try the problem with smaller or easier numbers.
- ✓ Set out your working clearly.
- ✓ Check your answer is sensible.
- ✓ Make sure you answer the question that was asked.

Worked example

2 Michael is organising a conference for 837 people.

Each person at the conference will be given a free pen.

The pens are sold in packs of 10 or 100

A pack of 10 pens cost £1.10

A pack of 100 pens cost £9.60

He wants to spend the smallest amount possible.

How much will he pay?

8 × £9.60 = £76.80

4 × £1.10 = £4.40

£76.80 + £4.40 = £81.20

Look at the combination of packs of pens that he needs to buy for 837 people.

He needs to buy:

8 packs of 100 pens = 800 pens and

4 packs of 10 pens = 40 pens

800 + 40 = 840 pens

He will have 3 pens left over.

Now try this

1 Joe buys a magazine that costs £4.45 and two birthday cards that cost £1.99 each. He pays with a £10 note. How much change will he receive?

2 Lulu is packing a suitcase for her holiday. She can pack up to 19 kg of luggage. Lulu's suitcase weighs 4.6 kg. She wants to pack 11.3 kg of clothing, 0.6 kg of toiletries and 1.1 kilograms of electrical equipment. Can she pack a guidebook weighing 0.7 kg as well? Explain your answer.

Fractions and decimals

You can compare and order fractions by writing them as decimals.

Writing a fraction as a decimal

You can think of fractions as the numerator divided by the denominator:

$$\frac{\text{numerator}}{\text{denominator}} = \text{numerator} \div \text{denominator}$$

To convert a fraction to a decimal, divide the numerator by the denominator.

Worked example

1 Write these fractions as decimals.

(a) $\frac{2}{25}$

$2 \div 25 = 0.08$

(b) $\frac{5}{16}$

$5 \div 16 = 0.3125$

Divide the numerator by the denominator. You can use your calculator to help you.

Worked example

2 The weights of some parcels are listed below.
Order the weights from smallest to largest.

$1\frac{2}{5}$ kg $2\frac{1}{3}$ kg $1\frac{3}{10}$ kg $2\frac{3}{5}$ kg

$1\frac{2}{5}$ kg $= 1.4$ kg $2\frac{1}{3}$ kg $= 2.333...$ kg

$1\frac{3}{10}$ kg $= 1.3$ kg $2\frac{3}{5}$ kg $= 2.6$ kg

From smallest to largest:

1.3 kg 1.4 kg 2.333... kg 2.6 kg

$1\frac{3}{10}$ kg $1\frac{2}{5}$ kg $2\frac{1}{3}$ kg $2\frac{3}{5}$ kg

First, write the fractions as decimals.

$1\frac{2}{5} = 1 + \frac{2}{5}$

$\frac{2}{5}$ can be written as $2 \div 5 = 0.4$

So $1\frac{2}{5} = 1 + 0.4$

$= 1.4$

Next, order the decimals from smallest to largest.

Finally, write the fractions in order.

Now try this

1 Write these fractions as decimals:

(a) $\frac{4}{5}$ **(b)** $\frac{7}{20}$ **(c)** $\frac{8}{40}$ **(d)** $1\frac{9}{12}$

2 Which fraction is larger, $\frac{3}{8}$ or $\frac{4}{9}$?

3 Order these fractions from smallest to largest: $\frac{5}{8}$ $\frac{6}{8}$ $\frac{4}{10}$ $\frac{7}{16}$

Percentages

Percentages are useful when comparing proportions of different amounts.

'Per cent' means 'out of 100'. A percentage can be written as a fraction with a denominator of 100.

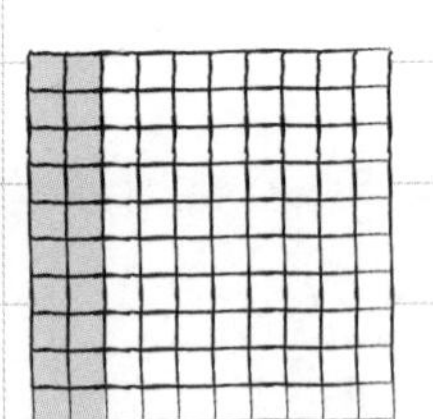

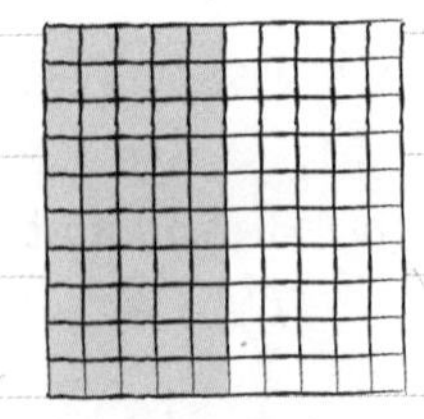

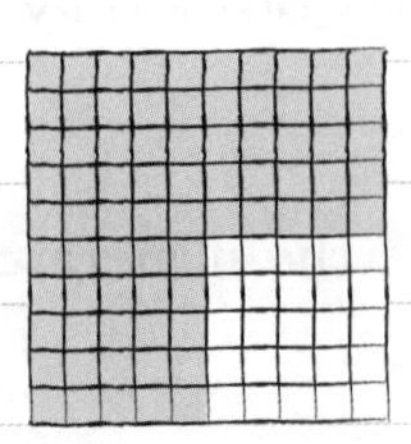

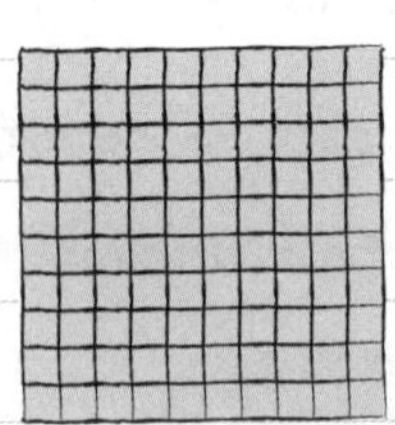

$20\% = \frac{20}{100}$ $50\% = \frac{50}{100}$ $75\% = \frac{75}{100}$ $100\% = \frac{100}{100}$

Worked example

Percentages of a whole always add up to 100%, so the sum of the percentage of students who pass and the percentage of students who fail must be 100%

1 At a driving school, 63% of students pass the driving test. The rest fail.
What percentage of students fail the test?

100% − 63% = 37%

Worked example

2 Each week, Carla spends 25% of her income on rent, 37% on food and $\frac{11}{100}$ on entertainment. She saves the rest.
What percentage of her income does Carla save?

25% + 37% + 11% = 73%

100% − 73% = 27%

Problem solved!

✓ Add together the percentages of her income that Carla spends on rent, food and entertainment. $\frac{11}{100}$ is the same as 11 out of 100 or 11%

✓ Subtract the percentage of Carla's income that she spends each week from 100% to find the amount she saves.

Now try this

1 Claire took a maths test. She got 32 questions wrong and didn't answer a further 7 questions. She got the rest right. There were 100 questions in total. She needed 60% or more to pass the test. Did Claire pass the test? Explain your reasons.

2 The table shows the number of calls made by a team of salespeople and the number of people who bought a product during one weekend. The salespeople made 100 calls in total.

	Saturday	Sunday
Number of calls	44	56
Number of sales	21	34

(a) What percentage of calls were made on Sunday?
(b) Over the whole weekend, what percentage of calls did not result in sales?

Calculating percentage parts

You will be asked to solve problems where you need to work out percentages of an amount.

Finding a percentage of an amount

To find a percentage of an amount:

1 Divide the percentage by 100

2 Multiply the result by your amount.

For example, 12% of 80 cm is 9.6 cm:

12 ÷ 100 = 0.12

0.12 × 80 cm = 9.6 cm

Worked example

1 Divya spends 24% of her wages on rent. She earns £2,000 per month. How much does she spend on rent?

24 ÷ 100 = 0.24

0.24 × £2,000 = £480

Worked example

2 Romesh books a holiday that costs £540 and pays a 15% deposit. How much has he left to pay?

15 ÷ 100 = 0.15

0.15 × 540 = £81

£540 − £81 = £459

Work out 15% of £540 then subtract this amount from £540 to work out what he still has left to pay.

Worked example

3 Jenny and Antonio are sorting eggs from their farm into boxes. They collect 130 eggs and 80% are large. The rest are medium.

(a) How many eggs are medium?

80 ÷ 100 = 0.8

0.8 × 130 = 104

130 – 104 = 26 so 26 eggs are medium.

(b) Large eggs cost 40p each. Medium eggs cost 70% of the cost of large eggs. How much do medium eggs cost?

70 ÷ 100 = 0.7

0.7 × 40p = 28p

Problem solved!

(a) ✓ Work out 80% of 130 eggs to find out how many are large.

✓ Subtract the number of large eggs from the total number of eggs to find out how many eggs are medium.

(b) Work out 70% of 40p. Remember to include the correct units in your answer.

Now try this

1 In a race, the £640 prize money is to be shared by the contestants with the four fastest times. The winner gets 50%, second place gets 30% and third place gets 15%. Fourth place gets the remaining prize money. How much does each of the four contestants get?

2 There are 750 students in Castle Hill School. 56% study French. The rest of the students study German. How many students study German?

Fractions, decimals and percentages

You can arrange a list of fractions, decimals and percentages in order of size by changing them to the same type.

Converting between fractions, decimals and percentages

1. You can convert a decimal to a percentage by multiplying by 100
   0.37 = 37%
2. You can write any percentage as a fraction with denominator 100
   $60\% = \frac{60}{100}$
3. You can convert a fraction to a decimal by dividing the numerator by the denominator.
   $\frac{3}{4} = 3 \div 4 = 0.75$

Useful equivalents

Remember these common fraction, decimal and percentage equivalents.

Fraction	Decimal	Percentage
$\frac{1}{100}$	0.01	1%
$\frac{1}{10}$	0.1	10%
$\frac{1}{5}$	0.2	20%
$\frac{1}{4}$	0.25	25%
$\frac{1}{2}$	0.5	50%
$\frac{3}{4}$	0.75	75%

Worked example

Thomasina has a bag of jelly beans.
15% of the jelly beans are strawberry flavoured.
$\frac{1}{4}$ of the jelly beans are pineapple flavoured.
$\frac{2}{5}$ of the jelly beans are apple flavoured.
The remaining jelly beans are cinnamon flavoured. What percentage of the jelly beans are cinnamon flavoured?

$\frac{1}{4} = 25\%$ $\frac{2}{5} = 40\%$
15 + 25 + 40 = 80 100 − 80 = 20
20% of the jelly beans are cinnamon flavoured.

Problem solved!

As the answer needs to be a percentage, convert the fractions to percentages.
All the jelly beans are 100% of the jelly beans. Work out the percentage of jelly beans that are **not** cinnamon flavoured.
Subtract your answer from 100% to find the percentage of jelly beans that are cinnamon flavoured.

Now try this

1 Write these numbers in order, starting with the smallest: 0.42 $\frac{2}{5}$ 36%

2 In a game, 5 players each put 12 counters on a table. Sasha wins 35% of the counters. Peter wins $\frac{2}{5}$ of the counters. Hayden wins the remaining counters. How many counters does Hayden win?

Write all three numbers in the same form, then compare them.

Word problems with percentages

When you are solving problems, you need to:

- read the question
- decide which calculation you are going to use
- check your answers
- make sure you have answered the question asked.

Worked example

1 Joseph owns a cafe. His customers can order a choice of two set meals.

One weekend, he has 80 customers. 20% of these customers order set meal A. The other customers order set meal B.

The table shows the cost of each set meal.

	Cost
Set meal A	£12
Set meal B	£15

How much money did Joseph's cafe take over the weekend?

set meal A:

$20 \div 100 = 0.2$

$0.2 \times 80 = 16$ people

$16 \times 12 = £192$

set meal B:

$80 - 16 = 64$ people

$64 \times 15 = £960$

total money earned:

$192 + 960 = £1,152$

Problem solved!

- Work out the number of people that ordered set meal A by finding 20% of 80
- Work out the number of people that ordered set meal B by subtracting the number of people who ordered set meal A from 80
- Work out the total money earned for set meal A and set meal B.
- Add the money earned from set meal A and set meal B to get the total money earned.

Worked example

2 A hotel did a survey to check customer satisfaction. In March, 6 out of 10 people said they found the services excellent.

In April, 3 out of 5 people said they found the services excellent.

Which month had the higher percentage of excellent responses?

March

$\frac{6}{10} = 6 \div 10 \times 100$

$= 60\%$

April

$\frac{3}{5} = 3 \div 5 \times 100$

$= 60\%$

They both had the same percentage of excellent responses.

Now try this

In one week, 150 people visited a sports centre. 30% used the gym, 20% used the swimming pool and the rest went to fitness classes. The cost of each of the activities is given in the table.

How much did the sports centre make in the week?

Activity	Cost
gym	£5
swimming pool	£6
fitness classes	£8

Using formulas

A formula is a mathematical rule that lets you calculate one quantity when you know the others.

Substituting values into formulas

To use a formula, substitute the values you know and then work out the answer. Remember to follow the correct order of operations.

Order of operations

Remember to use the correct order of operations when you are doing a calculation.

First, complete any calculations in brackets. Then do any division and multiplication.

Then do any addition and subtraction.

Worked example

1 The formula connecting speed to distance and time is: speed = distance ÷ time

Calculate the speed of a car that travels 200 km in 4 hours.

speed = 200 km ÷ 4 hours
= 50 km/h

Problem solved!

✓ Substitute the numbers into the formula.

✓ Check you have included the correct units.

✓ Check your answer makes sense.

Worked example

2 You can use this formula to work out the cooking time in minutes for a turkey:

cooking time = weight in kg × 30 + 45

Work out the cooking time for a turkey weighing 7 kg.

cooking time = 7 × 30 + 45
= 210 + 45
= 255 minutes

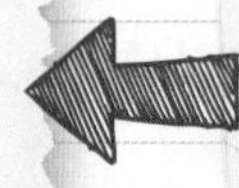

Substitute the weight into your formula before you do any calculations.

Remember to use the correct order of operations. You multiply before you add.

Now try this

Dogs age faster than humans. The formula for working out the age of a dog in dog years is:

dog years = 21 + 4.5 × human years

Find the age of a dog that has been living for 10 human years.

Ratio

Ratios are used to compare quantities.

Writing ratios

 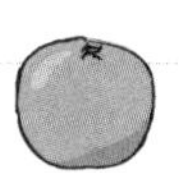  

Out of 4 pieces of fruit, there are 3 apples and 1 orange. The ratio of apples to oranges is 3 : 1

The sum of the two parts of this ratio is the number of items there are in total.

3 + 1 = 4

Equivalent ratios

You can find equivalent ratios by multiplying or dividing by the same number.

Simplest form

The simplest form of a ratio is the equivalent ratio with the smallest possible whole number values.

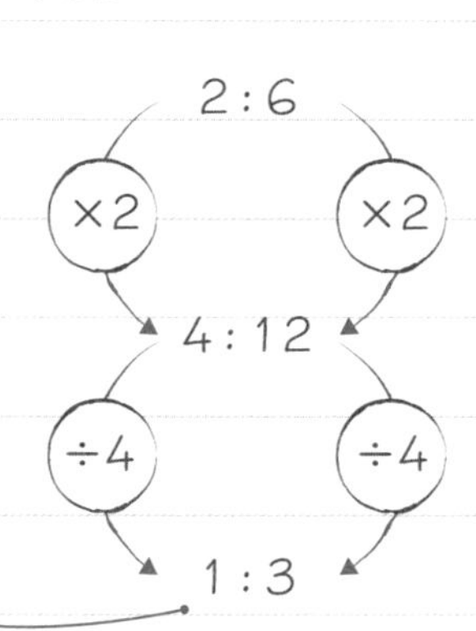

The ratio 1 : 3 is in its simplest form.

Worked example

Write each ratio in its simplest form.

(a) A garment label says 80% cotton 20% acrylic. What is the ratio of cotton to acrylic?

cotton : acrylic = 80 : 20
= 4 : 1

(b) A recipe uses 250 g of flour to 50 g of butter. What is the ratio of flour to butter?

flour : butter = 250 : 50
= 5 : 1

Write the ratios in the order they are given.
Simplify the ratios fully.
Divide 80 and 20 by 20
80 : 20 = 4 : 1
Divide 250 and 50 by 50
250 : 50 = 5 : 1
You can also simplify in stages.
80 : 20 = 8 : 2 = 4 : 1
250 : 50 = 25 : 5 = 5 : 1

Now try this

1 A bag contains green and red cherries in the ratio 1 : 4
Another bag contains green and red cherries in the ratio 3 : 15
Are these equivalent ratios? Explain your reasons.

2 Verity got 12 out of 16 in a test. She says that the ratio of correct to incorrect answers is 4 : 1. Explain why she is wrong.

Had a look ☐ Nearly there ☐ Nailed it! ☐

Ratio problems

You will need to use ratios to solve problems in different contexts.

Worked example

1 An electrician spent 30 minutes travelling to a client and 2 hours working for them.

Write the ratio of travel time to working time in its simplest form.

2 hours = 2 × 60 = 120 minutes

travel time : working time

30 : 120

1 : 4

You must make sure the ratios are in the same units before simplifying.

Either convert hours to minutes or minutes to hours.

Divide both parts of the ratio by 30 to get it in its simplest form.

Worked example

2 Jesse and Simone are buying a new computer.

They divide the cost in the ratio 1 : 3

Jesse pays £97.

How much more does Simone pay?

3 × £97 = £291

£291 − £97 = £194

Simone pays £194 more than Jesse.

Jesse's name comes first so the first number in the ratio refers to the amount he pays.

Jesse pays £97 and Simone pays three times as much. Multiply Jesse's share by 3 to find out how much Simone pays. Then subtract £97 from that amount to find the difference.

Now try this

1 To make meringue, John whisked 80 g of egg white with 240 g of sugar. What was the ratio of egg white to sugar? Write the ratio in its simplest form.

2 Tamal gives away a proportion of his wages to two charities. He donates money to a local food bank and a medical research charity in the ratio 3 : 1

Last year he gave away £990 to the food bank. How much did he give to the medical research charity?

Proportion

When two quantities are in direct proportion, they both increase or decrease at the same rate.

Money problems

Direct proportion problems often involve money. You can sometimes solve problems by working out the cost of **one item.**

If 3 theatre tickets cost £135:

1 theatre ticket costs £135 ÷ 3 = £45

9 theatre tickets cost 9 × £45 = £405

Worked example

1 Suresh buys 4 picture frames for a total cost of £11.40

Work out the cost of 7 of these picture frames.

Cost of 1 frame = £11.40 ÷ 4
= £2.85

Cost of 7 frames = £2.85 × 7
= £19.95

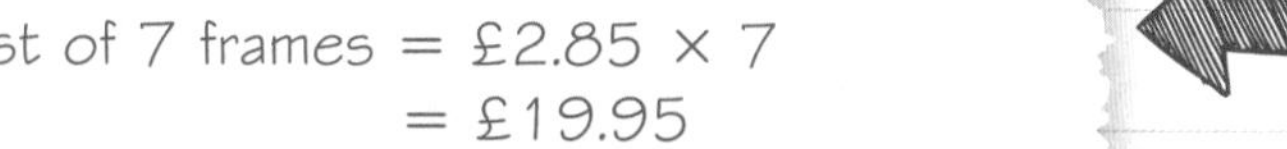

Calculate the cost of 1 picture frame first. Then multiply the cost of 1 frame by 7 to work out the cost of 7 frames. When you are working with money you should:

- do all your calculations in either pounds or pence
- write answers in pounds to 2 decimal places.

Worked example

2 A garden centre sells large and small plants.

(a) 15 large plants cost £75. How much will 11 large plants cost?

75 ÷ 15 = 5

11 × 5 = 55

11 large plants will cost £55

Work out how much 1 plant costs.
Multiply that amount by 11
Make sure you give units with your answer.

(b) Small plants are available in trays of 6 or boxes of 10

tray of 6	box of 10
£15	£24

Work out whether the box or the tray offers better value.

tray	box
15 ÷ 6 = 2.5	24 ÷ 10 = 2.4
£2.50 per plant	£2.40 per plant

£2.40 < £2.50 so the box is better value.

Work out the cost of each plant in a tray of 6 and in a box of 10. Write a short conclusion saying which one is better value.

Now try this

Lydia buys 8 identical bottles of water for a total cost of £4.48
Work out:

(a) the cost of 5 bottles of water.

(b) the cost of 12 bottles of water.

Recipes

A recipe will give the amount of ingredients for a certain number of people. Sometimes you will need to adjust the amount of ingredients to make enough for a different number of people. You can use direct proportion to calculate the correct quantity of ingredients.

Worked example

1 A recipe for biscuits uses 100 g of butter to make 8 biscuits.

How many grams of butter would you need to make 24 biscuits?

$24 \div 8 = 3$

$100\,g \times 3 = 300\,g$

Divide 24 by 8 to find out how many lots of the recipe you need to make.

Number of biscuits	Butter
8	100 g
24	300 g

×3 ×3

Worked example

2 This list of ingredients makes enough pizza to serve 6 people.

400 g bread dough

120 ml tomato puree

240 g grated cheese

Jonathan has 1200 g of bread dough, 300 ml of tomato puree and 800 g of grated cheese. Does he have enough ingredients to serve 18 people?

$18 \div 6 = 3$

Jonathan needs:

bread dough: $3 \times 400\,g = 1200\,g$

tomato puree: $3 \times 120\,ml = 360\,ml$

grated cheese: $3 \times 240\,g = 720\,g$

No, Jonathan does not have enough as he doesn't have enough tomato puree.

Divide the number of servings Jonathan needs by the number of servings the recipe makes to find out how many lots of the recipe he needs to make.

Work out the amount of ingredients for 18 people and compare to how much Jonathan has.

People	Bread dough	Tomato puree	Grated cheese
6	400 g	120 ml	240 g
18	1200 g	360 ml	720 g

×3 ×3

Now try this

Make sure you clearly state 'yes' or 'no' in your answer and explain your decision.

A recipe shows this list of ingredients to make vegetable soup for 4 people:
400 g of courgettes, 600 g of potatoes, 1000 ml stock

(a) How many grams of potatoes are needed to make the soup for 8 people?

(b) Simon has 1600 g of courgettes, 2000 g of potatoes, 5000 ml of stock. Does he have enough ingredients to serve 16 people?

Word problems with ratio

When solving problems, you need to make sure you have answered the question that has been asked. Always read the question carefully. Underlining the key information might help.

Worked example

1 A golf club has 200 members. 150 are female.

Work out the ratio of male to female members.

Give your answer in its simplest form.

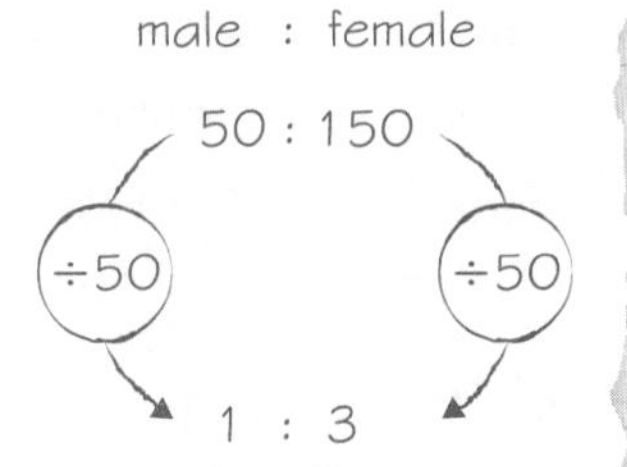

number of male members = total number of members − number of female members

200 − 150 = 50 male members

Problem solved!

- ✓ You need to write the ratio of males to females, so you need to find the number of males.
- ✓ Be careful with the order of the ratio.
- ✓ Check that the ratio has been simplified fully.

Worked example

2 Dan wants to buy some tiles for his kitchen.

He needs a total of 800 blue tiles. For every 4 blue tiles he will need 1 white tile.

Packs of 10 blue tiles cost £12 and packs of 10 white tiles cost £9

How much will it cost him to tile his kitchen?

800 ÷ 4 = 200

Dan needs 800 blue tiles and 200 white tiles.

cost of blue tiles

800 ÷ 10 = 80

80 × 12 = £960

cost of white tiles

200 ÷ 10 = 20

20 × 9 = £180

Total cost of tiles = £960 + £180
= £1,140

Problem solved!

- ✓ Work out how many white tiles he needs. Ratio of blue tiles to white tiles is 4 : 1

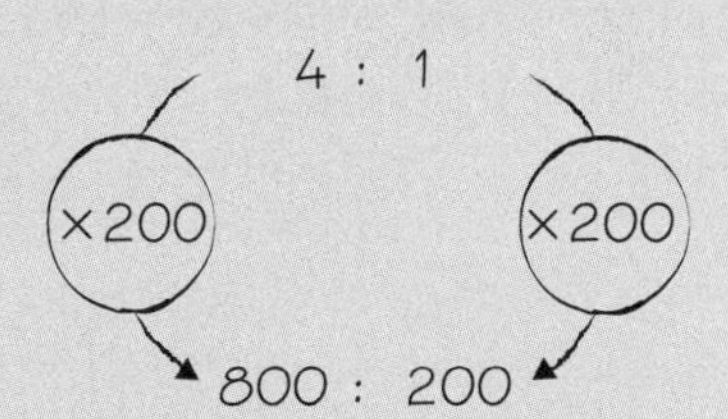

- ✓ He buys the blue tiles and white tiles in packs of 10. Work out the number of packs of blue tiles and white tiles he needs.
- ✓ Work out the cost of the blue tiles and white tiles.
- ✓ Find the total cost of the tiles.

Now try this

A caterer is buying ingredients to make tuna sandwiches.
For every 5 bread rolls, he needs 1 tin of tuna. He has 40 bread rolls.

(a) How many tins of tuna does he need?

(b) A packet of 10 bread rolls costs £4 and a pack of 2 tins of tuna costs £3
How much will it cost him to make the 40 sandwiches?

Problem-solving practice

When you are solving problems, you need to:

- ✓ read the question
- ✓ check your answers
- ✓ decide which calculation you are going to use
- ✓ make sure you have answered the question asked.

On a weekday, this is how much it costs to park on Barrington Road:

up to 2 hours	£3
2–4 hours	£8
more than 4 hours	£14

Amina needs to park her car for 3 hours on Wednesday, 1 hour on Thursday and 6 hours on Friday. She thinks it will cost less than £20. Is Amina correct?

(3 marks)

Word problems page 13

Read the question carefully.

Find out how much each day will cost Amina and add these amounts together.

TOP TIP

Make sure you answer the question that has been asked – to complete your answer you must write a statement to say whether or not Amina is correct.

Josie owns a furniture shop.

She buys 80 sofas to sell in her shop.
Josie pays £200 for each sofa.
She sells $\frac{3}{4}$ of the sofas for £300 each.
She then has a sale and sells the rest of the sofas for £250 each.
What is the total amount Josie makes?

(4 marks)

Word problems with fractions page 18

Work out how much Josie pays for the sofas. You need to work out how many sofas she sells for £300 and how many sofas she sells for £250

TOP TIP

Plan how to lay out your answer. You need to show what you are working out at each stage, so it might help to use headings.

Joe's cafe

coffee	£2.25
tea	£1.75
squash	99p
soup	£3.60
bread roll	80p
sandwich	£3.05

Zhara buys 4 coffees, 2 teas, 3 squashes, 4 soups, 5 sandwiches and 2 bread rolls. She pays with two £20 notes and a £10 note. How much change should she get?

(3 marks)

Word problems with decimals page 22

Work out the cost of 4 coffees, 2 teas, and so on. Then add up the totals to work out Zhara total bill. Work out how much Zhara gives the cashier. Finally, subtract to find her change.

TOP TIP

Remember to do all your calculations in either pounds or pence.

Problem-solving practice

In 2010, a business used 1800 reams of paper. In 2011, the same business used 20% more paper.

How many reams of paper did the business use altogether in 2011?

(4 marks)

Calculating percentage parts page 25

First work out 20% of 1800, then add this to 1800 to find out how much paper was used in total.

TOP TIP

Check if your answer is sensible.

A holiday company asked its customers to rate a hotel. 70% of customers said that the hotel was excellent. 1 out of 5 customers said the hotel was satisfactory.

The rest were unhappy with the hotel.

What percentage of customers were unhappy with the hotel?

(2 marks)

Fractions, decimals and percentages page 26

Convert 1 out of 5 to a percentage.

Add this to 70% and subtract the total from 100%.

Check your working. All the percentages should add up to 100%.

A flat-packed furniture company is packing door handles and screws for each of its cupboards.

For every cupboard, there need to be 4 screws and 2 door handles. There are 30 cupboards.

(a) How many screws are needed?

(1 mark)

(b) The cost price of 10 screws is £2 and the cost price of 5 door handles is £20

How much would the cost price be for the handles and screws of the 30 cupboards?

(3 marks)

Word problems with ratio page 33

(a) Notice that the cupboards and screws are in direct proportion.

(b)
- Work out the number of screws and door handles needed to make 30 cupboards.
- Work out the number of packs of screws and door handles needed.
- Work out the cost.

TOP TIP

In the online test, if you get stuck, you can flag a question to come back to later.

Click the review button so that you can check that question again.

Had a go ☐ Nearly there ☐ Nailed it! ☐

Problem-solving practice

Johanna needs to hire a car from Monday to Friday.

The formula below calculates the cost of hiring a car:

Cost (£) = £40 + number of days × £25

She has a budget of £150. Will she have enough money to hire the car?

(3 marks)

Using formulas page 28

- Work out the number of days Johanna needs to hire the car for.
- Substitute the numbers into the formula.
- Check your numbers are in the correct units.

Check whether or not your answer is sensible.

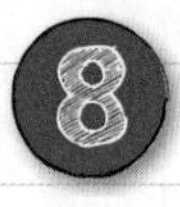

A shop sells packets of crisps. It has two offers.

offer A	offer B
12 packets for £4.32	18 packets for £6.30

Which offer is the best value for money?

(3 marks)

Proportion page 31

- Work out the cost of each packet of crisps for offer A.
- Work out the cost of each packet of crisps for offer B.
- Write a short conclusion saying which one is better value.

TOP TIP

Make sure you compare both costs in the same units. You can either use pounds or pence.

This recipe shows the list of ingredients to make scones for 8 people:

Ingredients
240 g self-raising flour
1 tsp baking powder
80 g butter
160 ml milk

Claire has:

500 g self-raising flour
3 tsp baking powder
200 g butter
300 ml milk

Does she have enough ingredients to make scones for 16 people?

(3 marks)

Recipes page 32

Work out the amount of each ingredient Claire will need to make scones for 16 people.

Use the onscreen calculator to do your calculations.
Show your working clearly so that you can answer the question with reasons.

Units of time

You need to be able to use time measured in different units.

Make sure you know these facts about time:

1 minute = 60 seconds	1 year = 365 days	1 day = 24 hours	1 decade = 10 years
1 hour = 60 minutes	1 year = 52 weeks	1 week = 7 days	1 century = 100 years

Worked example

1 (a) How many seconds are in 3 minutes?

$3 \times 60 = 180$ seconds

(b) How many hours in 420 minutes?

$420 \div 60 = 7$ hours

To convert from a bigger unit to a smaller unit, you need to multiply.

To convert from a smaller unit to a bigger unit, you need to divide.

(a) To convert from minutes to seconds, multiply by 60

(b) To convert from minutes to hours, divide by 60

Worked example

2 The table shows the amount of time Karen spent on certain tasks at work.
What was the total time she spent on these three tasks?

Task	Time
filing	2 hours
typing	90 minutes
answering the phone	2.5 hours

method 1

typing: $90 \div 60 = 1.5$ hours

$2 + 1.5 + 2.5 = 6$ hours

method 2

filing: $2 \times 60 = 120$ minutes

answering the phone: $2.5 \times 60 = 150$ minutes

$120 + 90 + 150 = 360$ minutes

Problem solved!

There are different ways you can answer this question, but you need to convert the times to the same units.

Method 1

Convert all times to hours.

Method 2

Convert all times to minutes.

You can convert 360 minutes to hours by dividing by 60

$360 \div 60 = 6$ hours

Now try this

1 How many minutes are there in 4 hours?

2 James spends 4.5 hours working and 30 minutes travelling to work. What is the total time James spends travelling and working?

Had a look ☐ Nearly there ☐ Nailed it! ☐

Dates

You need to be able to work out the number of days between given dates and use a calendar.

Days of the year

Month	Number of days
January	31
February	28 (29 in a leap year)
March	31
April	30
May	31
June	30
July	31
August	31
September	30
October	31
November	30
December	31

A leap year happens every 4 years. If the year is a multiple of 4, it will be a leap year. 2016 was a leap year.

Worked example

1 This calendar shows the month of October in 2015.

October

Mon	Tue	Wed	Thu	Fri	Sat	Sun
			1	2	3	4
5	6	7	8	9	10	11
12	13	14	15	16	17	18
19	20	21	22	23	24	25
26	27	28	29	30	31	

(a) How many days are there from 12 October to 27 October?

15 days

(b) Simon goes on holiday for 5 nights. He leaves on 29 October. What day does he return?

3 November

Count on 5 nights from 29 October.

Worked example

2 Sam ordered a cake on Monday 27 April 2015

The bakery said the cake would be ready on the Wednesday of next week.

What date was the cake ready?

Mon 27 Apr, Tue 28 Apr, Wed 29 Apr, Thu 30 Apr, Fri 1 May, Sat 2 May, Sun 3 May, Mon 4 May, Tue 5 May, Wed 6 May

The cake was ready on Wed 6 May.

There are 30 days in April.
List the days and their dates until Wednesday of the following week.

Now try this

1 Look at the calendar showing October 2015 above.

(a) How many days are there from Sunday 4 October to Wednesday 21 October?

(b) Thierry books a hotel from 9 October for 10 nights. What day does he leave the hotel?

2 A customer at a garage books his car in on Friday 29 May. He is told the car will be ready on the Thursday of the following week. What is the date of Thursday the following week?

12-hour and 24-hour clocks

Time can be displayed using the 12-hour clock or the 24-hour clock.

Reading a 12-hour clock

This clock tells the time in a train station one afternoon. The time it shows is 15 minutes past one or quarter past one.

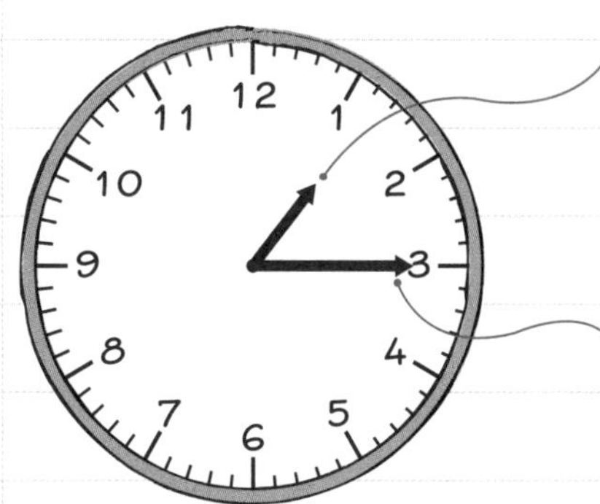

In the 12-hour clock, this is written as 1.15 p.m.

In the 24-hour clock, this is written as 13:15

Morning or afternoon

In the 12-hour clock you have to write whether it is a.m. (morning) or p.m. (afternoon).

In the 24-hour clock you can tell whether the time is afternoon or morning by looking at the first two digits.

If the first two digits are greater than 12 the time will be afternoon; if the first two digits are less than 12 the time will be morning.

Converting between clocks

12-hour clock	24-hour clock
8.15 a.m.	08:15
4.50 p.m.	16:50
12.00 p.m. (midday)	12:00
12.00 a.m. (midnight)	00:00

Worked example

A chef puts a piece of meat in the oven at 11.45 a.m.

It needs to be cooked for 2 hours.

What time does the chef take the meat out of the oven? Write the time in the 24-hour clock.

13:45

Add 2 hours onto 11.45 a.m.

The time will be in the afternoon so the hours will be a number larger than 12

Now try this

1 This clock shows a time in the afternoon.

(a) Write this time in the 12-hour clock.

(b) Write this time in the 24-hour clock.

2 Complete this table of times in the 12-hour and 24-hour clock.

12-hour	24-hour
3.12 a.m.	……
……	19:50
11.30 p.m.	……
……	10:15

Had a look ☐ Nearly there ☐ Nailed it! ☐

Timetables

Reading timetables is a skill that you will use when travelling. You need to know how to read timetables in order to work out how long journeys will take.

Timetables

Here is a bus timetable:

Crook	08:15	09:15	10:45	11:15
Prudhoe	08:28	09:28	10:58	11:28
Hexham	08:45	09:45	11:15	11:45
Alton	09:00	10:00	11:30	12:00

This bus leaves Crook at 10:45 and arrives in Hexham at 11:15

This bus leaves Prudhoe at 11:28 and arrives in Alton at 12:00

11:28 to 11:30 = 2 minutes
11:30 to 12:00 = 30 minutes
The journey time is 32 minutes.

Worked example

This is part of a train timetable from Richmond to Acton Central.

Richmond	1108	1128	1138
Kew Gardens	1111	1131	1141
Gunnersbury	1114	1134	1144
South Acton	1117	1137	1147
Acton Central	1120	1140	1150

(a) If Joseph needs to get to Acton Central before 11:45, which is the latest train he could get from Richmond?

11:28 from Richmond

Look at the row for South Acton and find the last train that arrives before 11:45

(b) How long does it take to get from Kew Gardens to Acton Central?

9 minutes

Find the arrival time and departure time in the table, then work out how far apart they are in minutes.

Now try this

This timetable shows the train times from Watford Junction to Euston.

(a) Mike gets on the 05:41 train from Watford Junction. What time does he arrive at Harlesden?

(b) If Mike takes the 05:11 train from Watford Junction, how long does the journey take to Euston?

~~50 minutes~~
45 minutes

Watford Junction	0511	0541	0611
Carpenders Park	0519	0549	0619
Harrow & Wealdstone	0527	0557	0627
North Wembley	0533	0603	0633
Harlesden	0540	0610	0640
Queen's Park	0547	0617	0647
Euston	0559	0631	0658

Creating a time plan

You need to know how to create a time plan. Start by adding in any activities that need to happen at a particular time. Then add the other activities in, making sure there is enough time for each one.

Worked example

1 Josie is planning some activities at an outdoor activity centre.

There are four activities: swimming zip wire treasure hunt assault course

- There are three groups of people.
- Every group must do each activity once.
- Only one group can do an activity in a session.

All the groups must do swimming in session two, three or four.

Plan the day for Josie.

	Session 1	Session 2	Session 3	Session 4
Group A	zip wire	swimming	treasure hunt	assault course
Group B	assault course	zip wire	swimming	treasure hunt
Group C	treasure hunt	assault course	zip wire	swimming

Start by adding in the swimming sessions. Then add in the other sessions for each group, making sure that no two groups are doing the same activity at the same time.

Worked example

2 Mike is planning a badminton tournament.

There are three players – A, B and C. Each player must play every other player once.

- The first game starts at 9.00 a.m.
- Each game lasts 40 minutes.
- There is a break of 10 minutes after each game.
- Only one game is played at a time.

Plan a timetable for the tournament showing the start time and players for each game.

Write down the start time and end time for each activity, making sure they don't overlap.

9.00–9.40	A plays B
9.40–9.50	Break
9.50–10.30	A plays C
10.30–10.40	Break
10.40–11.20	B plays C

Now try this

James is organising a group visit to a theme park. The group will arrive at 10.00 a.m. and leave at 4.00 p.m. Here is a list of all of the activities they will need to fit into their day.

Design a time plan for the day that shows the start and finish time for each activity.

wildlife show 12.00 p.m.–1.30 p.m.
adventure rides (1 hour)
treasure hunt (30 minutes)
nature tour (1 hour)
bird sanctuary (30 minutes)
lunch (30 minutes)
arts and crafts (1 hour)

Had a go ☐ Nearly there ☐ Nailed it! ☐

Problem-solving practice

When you are solving problems, you need to:

- ✓ read the question
- ✓ check your answers
- ✓ decide which calculation you are going to use
- ✓ make sure you have answered the question asked.

The table shows the amount of time it takes for some services to be completed by a mechanic.

Task	Time
car service	1 hour
tyre change	30 minutes
engine oil change	15 minutes
diagnostic check	45 minutes

In one day, a mechanic does 2 car services, 4 tyre changes, 3 engine oil changes, 1 diagnostic check.

How long do these tasks take in total? Give your answer in hours and minutes.

(3 marks)

Units of time page 37

Change the units of time so that you are working in minutes. At the end of your calculations, convert the time to hours and minutes.

TOP TIP

If your answer is a decimal, the numbers before the decimal point are the hours, but the numbers after the decimal point are a fraction of an hour, not the number of minutes.

A bride orders a wedding dress on Monday 21 September. The dressmaker needs 12 working days to make the dress and she does not work at weekends.

Will the bride get it in time for her wedding on Thursday 8 October?

(3 marks)

Dates page 38

Write out the dates from 21 September to 8 October and count how many working days there are.

TOP TIP

Make sure you answer the question that has been asked. Write a statement to say whether the bride will get her dress in time for the wedding.

Edward starts work at 8.30 a.m.

He works for 2 hours and then has a break for 30 minutes. He works a further $2\frac{1}{2}$ hours and then has lunch for 1 hour. He then works for another 2 hours.

What time does Edward finish work? Write your answer in the 24-hour clock.

(3 marks)

12-hour and 24-hour clocks page 39

Draw a table to help with the timings of his day.

starts work	08:30
start of break	
back at work	
start of lunch	
back at work	
ends work	

TOP TIP

As your answer is needed in the 24-hour clock, write the times in the table in the 24-hour clock.

Problem-solving practice

Here is part of a railway timetable.

Bath	0815	0845	0915
Chippenham	0830	0900	0930
Swindon	0850	0920	0950
Didcot	0915	0945	1015
Reading	0935	1005	1035

Jamelia gets to the station in Chippenham at 08:45. She waits for the next train to Didcot.

(a) How long does she have to wait?

(1 mark)

(b) At what time should she arrive at Didcot?

(1 mark)

All the trains take the same time to travel from Bath to Reading.

(c) How long does it take in minutes to travel from Bath to Reading?

(2 marks)

Timetables page 40

Look on the timetable for the next time after 08:45 that the train leaves Chippenham.

Look at the same column in the timetable to find out when that train arrives at Didcot.

In the online test, if you get stuck, you can flag a question to come back to later.

Flag

Click the review button so that you can check that question again.

Carl is organising an interview day for a candidate.

He will arrive at 10.00 a.m. and will leave at 2.00 p.m.

The list shows the activities that he will be doing during the day.

- interview (1 hour)
- data task (30 minutes)
- tour around site (45 minutes)
- inbox task (45 minutes)
- test (15 minutes)
- lunch (45 minutes)

The interview will be the last activity before the candidate leaves.

Design a time plan for the day.

The time plan must show the start and finish time for each activity.

(3 marks)

Creating a time plan page 41

Draw a table and include the start and finish times. Make sure you have enough room to fit the six different activities. Write in any activities that need to be done at a certain time first.

Time	Activity
10:00–	activity 1
	activity 2
	activity 3
	activity 4
	activity 5
13:00–14:00	interview

TOP TIP

The interview is at the end of the day so you can write this in first before completing the rest of the table.

Units

Most of the units of measurement used in the UK are metric units such as litres, kilograms and kilometres. You need to be able to estimate and know which units are appropriate.

Choosing units

You use different units when measuring different things.

If you are measuring weight, you will use grams, kilograms, or tonnes.

If you are measuring distance, you will use millimetres, centimetres, or kilometres.

If you are measuring capacity, you will use millilitres, centilitres or litres.

The word 'kilo' comes from a Greek word meaning 1000

Worked example

1 Write the units you would use to measure each of the following:

(a) The distance between two towns.

kilometres

(b) The height of a garden shed.

metres

(c) The weight of a bag of potatoes.

grams or kilograms

(d) The capacity of a car fuel tank.

litres

Worked example

2 Estimate each of these in metric units:

(a) The height of an average male in the UK.

about 1.8 m or 180 cm

(b) The weight of a cat.

about 4.5 kg

(c) The capacity of a coffee mug.

about 250 ml

3 The height of this man is 2 m. Estimate the height of the buiding.

$2 \times 4 = 8$

The building is about 8 m tall.

The building is about 4 times the height of the man.

Now try this

1. Bob measured the weight of an apple and said it was about 300 cm. Is Bob correct? Explain your reasons.
2. Which units would you use to measure the width of an adult thumb?
3. Which units would you use to measure the weight of a car?
4. Which units would you use to measure the capacity of a household boiler?

Measuring lines

You need to be able to use a ruler to draw and measure straight lines accurately. Don't measure lines in your exam unless the question tells you that the diagram is accurate.

Worked example

Here is a line AB.

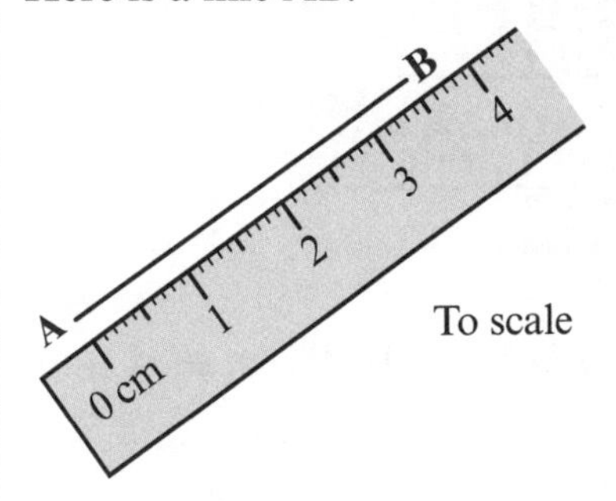

To scale

Measure the length of the line AB. 3.5 cm

Problem solved!

Line up the 0 mark on your ruler carefully with the start of the line at A.

Always measure to the nearest millimetre.

Make sure your ruler doesn't move while you're measuring the line.

Always write the units with your answer.

Drawing lines checklist

- ✓ Check whether you are working in cm or mm.
- ✓ Start the line at the 0 mark on your ruler.
- ✓ Hold your ruler firmly.
- ✓ Use a sharp pencil.
- ✓ Draw to the nearest mm.
- ✓ Label the length you have drawn.

Estimating

You can use lengths that you know to estimate other lengths.

This diagram shows a man standing next to a building.
The man is 3 cm tall in the drawing and the building is 12 cm tall.
The building is four times as tall as the man.

A good estimate for the height of an adult male is 1.8 m.

$4 \times 1.8\text{m} = 7.2\text{m}$

A good estimate for the height of the building is 7.2 m.

Now try this

1 Measure the lengths of these lines.

(a) ____________

(b) ____________________

(c) __________________

(d) ______________________

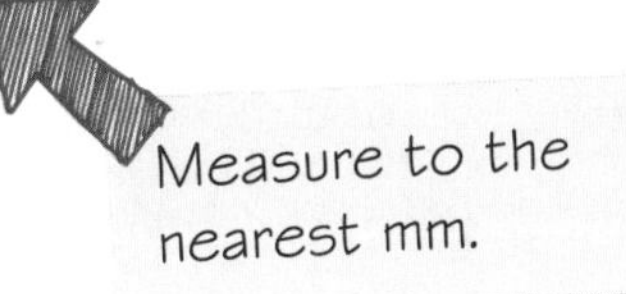

2 The picture of a surfboard and a London Bus have been drawn accurately to the same scale. The real length of the surfboard is 2 m. Estimate the real length of the London Bus.

Scales

You need to be able to accurately read scales and number lines and make estimates of readings.

Reading a scale

Begin by working out what each division on a scale represents.

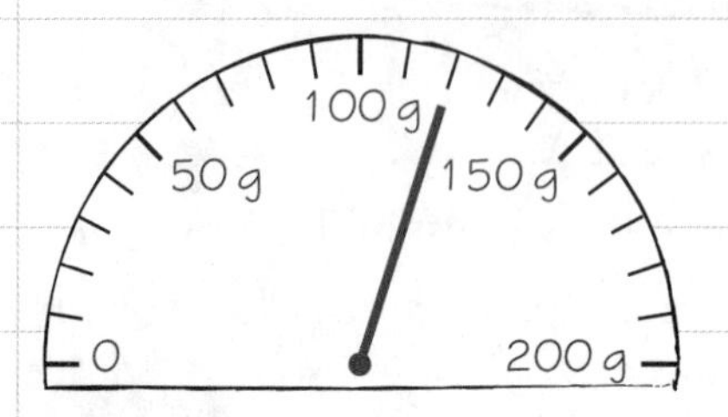

There are 5 divisions between 100 g and 150 g.

150 – 100 = 50

50 ÷ 5 = 10

Each division represents 10 g.

The scale reads 120 g.

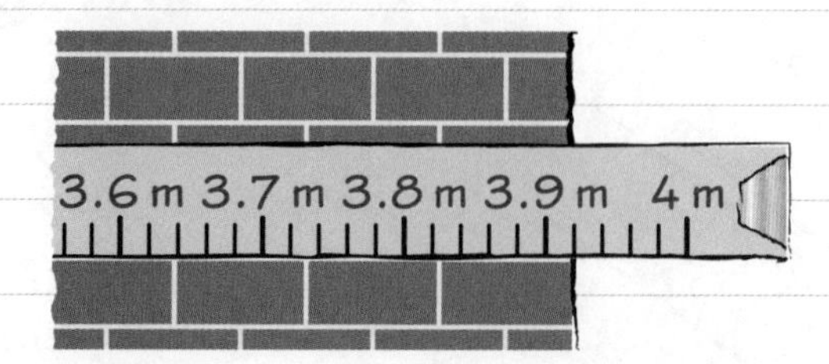

There are 5 divisions between 3.9 m and 4 m.

4 – 3.9 = 0.1

0.1 ÷ 5 = 0.02

Each division represents 0.02 m.

This wall is 3.92 m long.

Worked example

On the number line, mark with an arrow the number 8.8

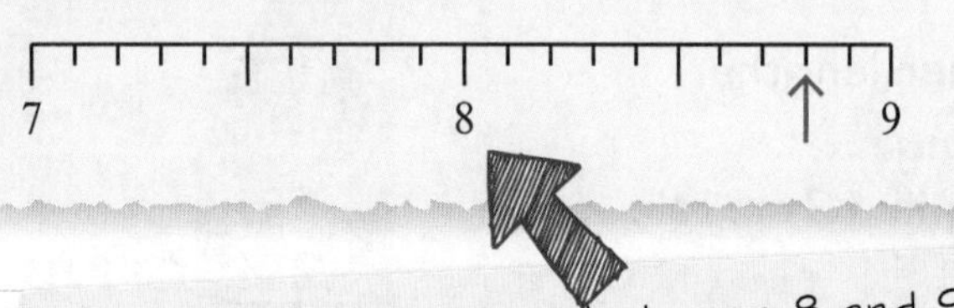

There are 10 divisions between 8 and 9 so each division represents 0.1

Estimating a scale reading

Sometimes, you might have to estimate the reading on a scale.

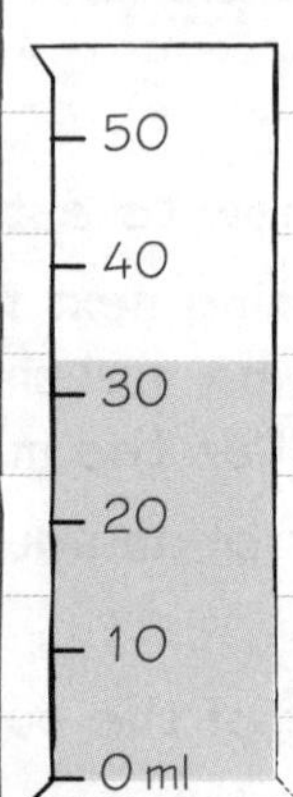

The water doesn't come up to an exact mark but you can make an estimate.

The water is closer to 30 ml than 40 ml.

33 ml would be a good estimate.

Now try this

(a) Which arrow marks the number 48 on the scale?

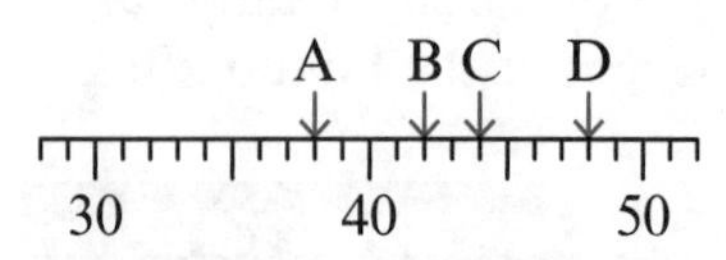

(b) Which arrow marks the number 637 on the scale?

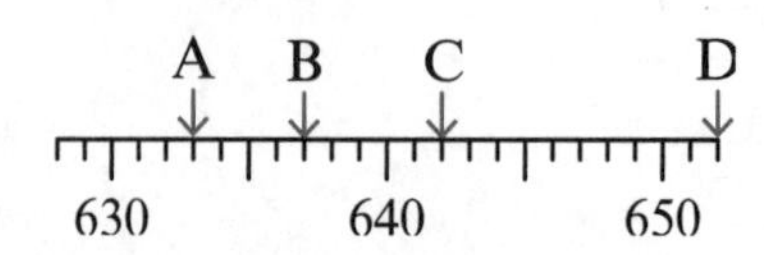

Mileage charts

You can use mileage charts to show the distance between two places. Distances in mileage charts can be given in miles or kilometres.

Worked example

1 Use the mileage chart to find the distance from Manchester to Worcester. The distances are given in miles.

Read down the column for Manchester and along the row for Worcester. The distance is in the right column **and** the right row: 99 miles.

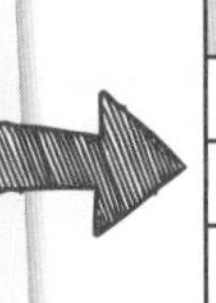

Manchester			
69	Wolverhampton		
99	31	Worcester	
64	122	155	York

The distance from Manchester to Worcester is 99 miles.

Worked example

To find the closest city, read across the row and find the shortest distance.

2 The mileage chart shows the distance between different cities given in miles.

(a) Which of these cities is closest to London?

Brighton

(b) How many miles is Bristol from London?

117 miles

Brighton			
134	Bristol		
183	124	Derby	
52	117	125	London

Other mileage charts

Sometimes you will see mileage charts that look like this.

	Aberystwyth	Beighton	Chard	Hertford
Aberystwyth		169	168	218
Beighton	169		214	135
Chard	168	214		163
Hertford	218	135	163	

There are different ways to find the distances between the towns.

To find the distance between Beighton and Chard you can:

- read across the row from Beighton to Chard
- read down the column from Beighton to Chard.

Now try this

The mileage chart shows the distances between cities in France. All distances are given in kilometres.

(a) Which is the closest city to Paris?

(b) What is the distance between Chartres and Perpignan?

Chartres			
362	Grenoble		
53	340	Paris	
492	275	513	Perpignan

Routes

Route maps can be useful for working out shortest routes.

Worked example

The diagram shows the distances between visitor attractions at Tumbledown Castle.

cafe
412 m
300 m
castle
220 m
gardens
140 m
180 m
252 m
lake

(a) What is the distance between the castle and the lake?

180 m

(b) Which attraction is 252 m from the lake?

the gardens

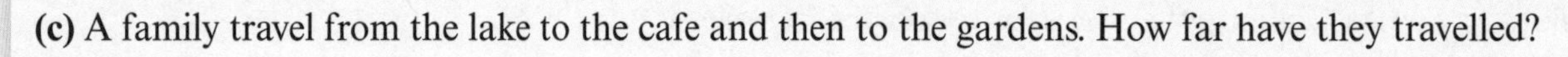

(c) A family travel from the lake to the cafe and then to the gardens. How far have they travelled?

140 + 300 = 440 m

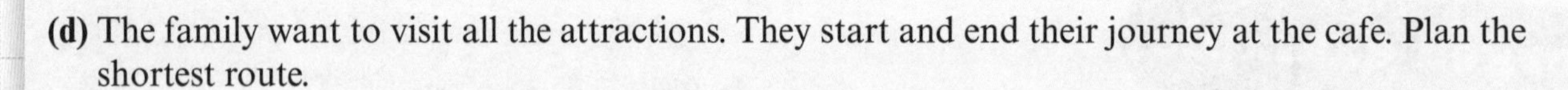

(d) The family want to visit all the attractions. They start and end their journey at the cafe. Plan the shortest route.

Start by going from the cafe to the lake, then to the castle, then to the gardens, then back to the cafe.

140 m + 180 m + 220 m + 300 m = 840 m

Plan your route to avoid walking between the cafe and the castle as this is the longest distance.

Problem solved!

When you are trying to find the shortest route, work through all of the options in turn.

Add together each distance to check the total length of the route.

Now try this

The diagram shows the distances between four popular destinations.

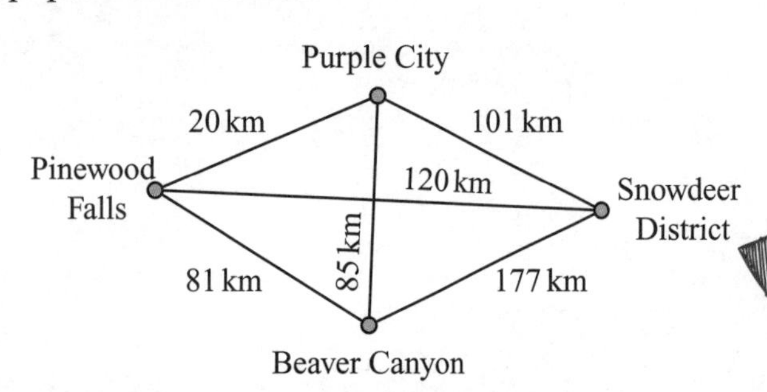

(a) What is the distance from Beaver Canyon to Pinewood Falls?

(b) Jacquie is organising a group holiday. She wants to visit all four destinations.

She starts and finishes her journey in Purple City. Plan the shortest route that she can travel.

Try different journeys until you are confident you have found the shortest one.

Length

You need to be able to recognise units of length and convert between different units. Metric units of length include kilometres (km), metres (m), centimetres (cm) and millimetres (mm).

For a reminder about the different units of length, see page 44.

Converting between metric units of length

You can convert between metric units of length by multiplying or dividing by 10, 100 or 1000.

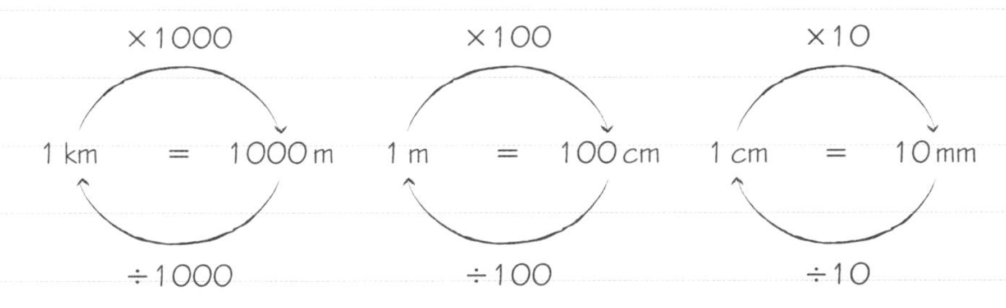

Worked example

1 (a) Convert 3 kilometres (km) to metres (m).

——→ ×	
km	m
1	1000
3	3000

3 × 1000
= 3000 m

Draw a table and write down what one unit is equal to. This number tells you what you need to multiply or divide by.

To change from km to m, multiply by 1000

To change from m to km, divide by 1000

(b) Convert 5000 metres (m) to kilometres (km).

←—— ÷	
km	m
1	1000
5	3000

5000 ÷ 1000
= 5 m

Worked example

2 Order these units of length from smallest to largest:

40 000 cm 4 km 40 m

Before you start, convert all units to the same unit. You can choose either km, cm or m.

In the example, the units chosen are m.

40 000 cm = 400 m

4 km = 4000 m

40 m < 40 000 cm < 4 km

Now try this

1 Convert: **(a)** 7 m to cm. **(b)** 300 cm to m. **(c)** 2.5 km to m. **(d)** 12 000 m to km.

2 Order these lengths from smallest to largest.
6000 mm 60 cm 60 m

Weight

You need to be able to recognise units of weight and convert between different units. Common metric units of weight include: tonnes (t), kilograms (kg), grams (g) and milligrams (mg).

Converting between metric units of weight

You can convert between metric units of weight by multiplying or dividing by 1000

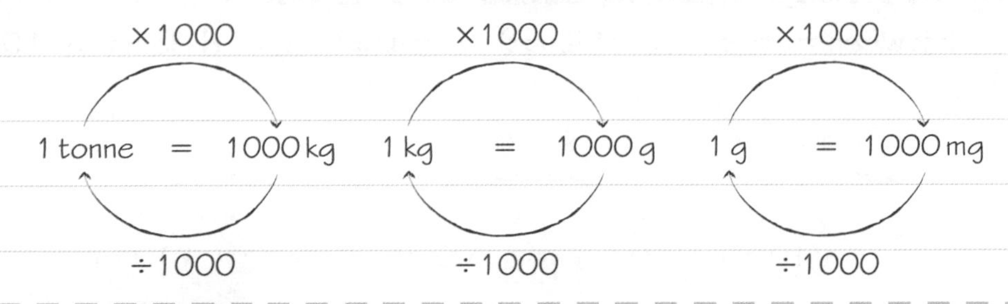

Worked example

1 (a) Convert 8 kg to g.

→ ×	
km	m
1	1000
8	8000

8 × 1000
= 8000 g

(b) Convert 9000 mg to g.

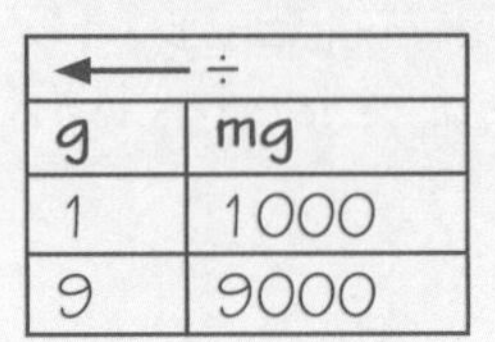

← ÷	
g	mg
1	1000
9	9000

9000 ÷ 1000
= 9 g

Draw a table and write down what one unit represents. This number tells you what you need to multiply or divide by.
To change from kg to g, multiply by 1000
To change from mg to g, divide by 1000

Worked example

2 Here are the weights of three boxes:

2 kg, 1000 g, 5000 mg.

Work out the total weight.

2 kg = 2000 g
5000 mg = 5 g

total weight = 2000 + 1000 + 5
= 3005 g

Before you start, convert all units to the same unit.
You can choose either kg, g or mg.
In the example, the units chosen are g.

Now try this

1 Convert: **(a)** 3 g to mg **(b)** 6000 g to kg **(c)** 6.2 kg to g.

2 Find the total of these weights: 8 kg 4000 g 30 000 mg

Capacity

Capacity tells you how much a container can hold. You need to be able to recognise units of capacity and convert between different units. Metric units of capacity include litres (l), centilitres (cl) and millilitres (ml).

Converting between metric units of capacity

You can convert between metric units of capacity by multiplying or dividing by 10, 100 or 1000

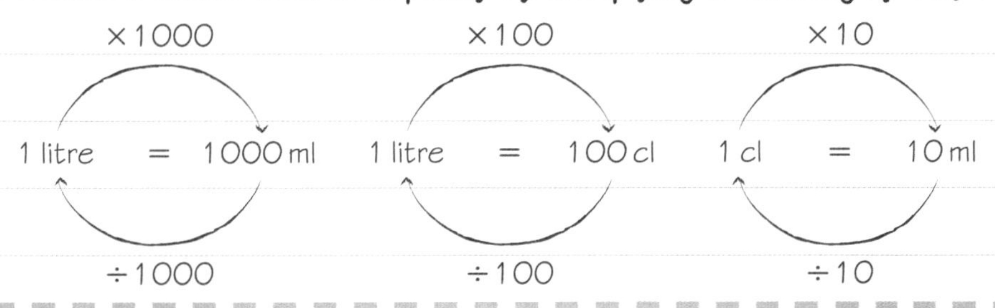

Worked example

1 (a) Convert 9 litres to millilitres.

→ ×	
litres	millilitres
1	1000
9	9000

$9 \times 1000 = 9000$ ml

(b) Convert 40 millilitres to centilitres.

← ÷	
centilitres	millilitres
1	10
4	40

$40 \div 10 = 4$ cl

Check that your answer is sensible. If 1 cl is 10 ml, then 4 cl will be more than 10 ml.

Problem solved!

✓ Draw a table and write down what one unit represents.

✓ This number tells you what you need to multiply or divide by.

- To change from litres to millilitres, multiply by 1000
- To change from millilitres to centilitres, divide by 10

Worked example

2 A smoothie is made from 150 ml of milk.
A shop has a 3 litre bottle of milk.
Can it make 25 smoothies?

3 litres = 3000 millilitres

3000 ml ÷ 150 ml = 20 smoothies

No, the shop doesn't have enough milk to make 25 smoothies.

Before you start, convert all units to the same unit. You can choose either ml or litres.

Now try this

1 Convert 5 litres to millilitres.

2 Brian needs 5 litres of paint to paint a wall. Paint is sold in 2 litre tins and 500 millilitre tins.
Brian wants to buy as few tins as possible and doesn't want any paint left over.
What combination of tins of paint does he need to buy?

Money

You need to be able to calculate with money and convert between pounds and pence.

Converting between pounds and pence

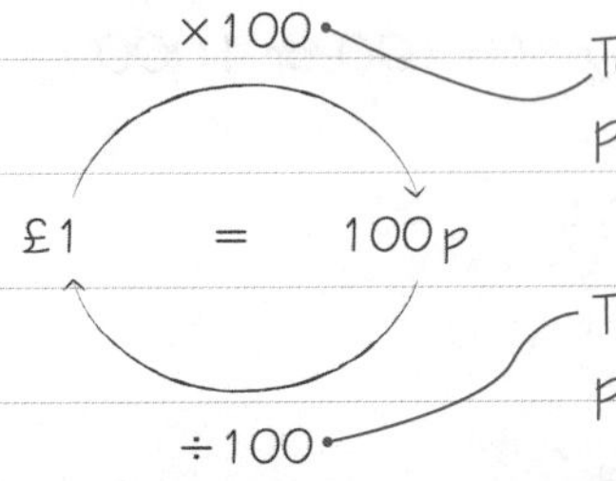

To convert pence to pounds, divide by 100

To convert pounds to pence, multiply by 100

Worked example

1 (a) What is £23.45 in pence?

£23.45 × 100 = 2345p

(b) What is 85p in pounds?

85p ÷ 100 = £0.85

Worked example

2 Caitlin buys a burger for £2.30, a cup of coffee for 99p and a cake for £1.20.

(a) How much is the bill?

99p ÷ 100 = £0.99

£2.30 + £0.99 + £1.20 = £4.49

(b) How much change does she get from a £10 note?

£10 − £4.49 = £5.51

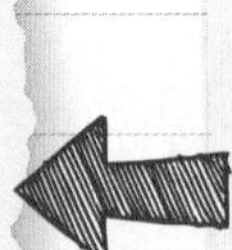

Convert all of the amounts so that they are in the same units: pounds or pence.

Change the price of the coffee from pence to pounds by dividing by 100

When all the prices are in pounds you can add them up to find the total.

Worked example

3 A supermarket sells bags of crisps individually or in packs of 6.

An individual bag of crisps costs 50p

A pack of 6 bags of crisps costs £2.70

Which of these options is the best value for money?

6 pack: price of 1 bag = £2.70 ÷ 6

= £0.45

The 6 pack is better value for money.

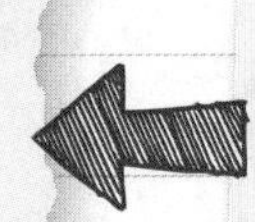

You can compare costs by finding the cost of one of the items.

To work out the cost of one item work out:

cost of all the items ÷ number of items

Now try this

1 Emma and her two children are going on a trip. The cost of an adult bus ticket is £2.60 and the cost of a child bus ticket is 75p. How much does the bus ride cost them in total?

2 A shop sells pens in packs of 3 or 5

A pack of 3 pens cost £1.26 and a pack of 5 pens cost £1.95

Which is the better value?

Remember to answer the question. Decide which is better value for money and use your calculations to show your reasons.

Temperature

You will usually see temperature measured in degrees Celsius (°C).

You need to be able to read temperature scales and work out the differences between temperatures.

Thermometers

A thermometer measures temperature.

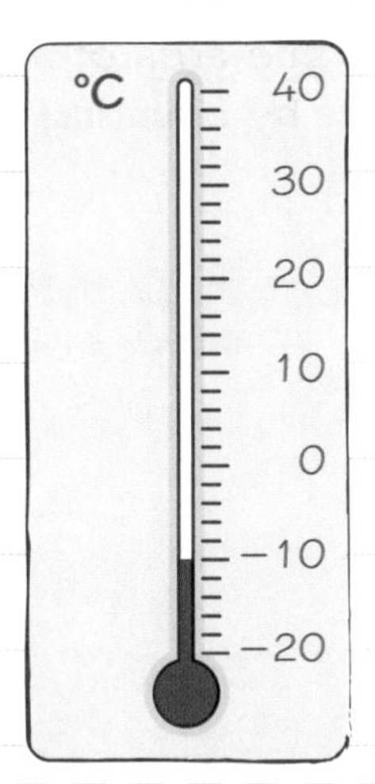

This thermometer shows the temperature in °C.
The temperature reading is −10°C.
The freezing temperature of water is 0°C.
The boiling temperature of water is 100°C.

Worked example

Go to page 46 for more help reading scales.

The normal temperature of an adult is between 36.5 °C and 37.2 °C.

This thermometer shows the temperature of a patient.

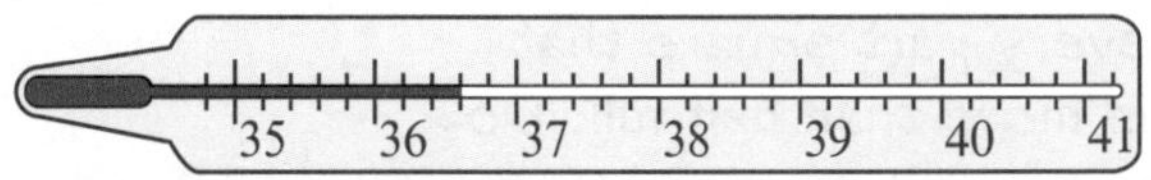

(a) What temperature does the thermometer show?

36.6 °C

(b) Is it within the normal range for an adult?

Yes

Work out what each mark on the thermometer scale represents.

Every large marker represents 1 °C.

There are 5 smaller markers between every large marker. Each small marker represents 1 °C ÷ 5 = 0.2 °C.

Count the number of marks from 36 and add on 0.2 °C for each mark as you go along.

Now try this

Food in a restaurant freezer must be kept at –18 °C or below.

A chef reads the temperature on three freezers in his kitchen.

(a) What is the temperature of each of the freezers?

(b) Are the freezers at the correct temperature? Give your answer and explain your reasons.

A

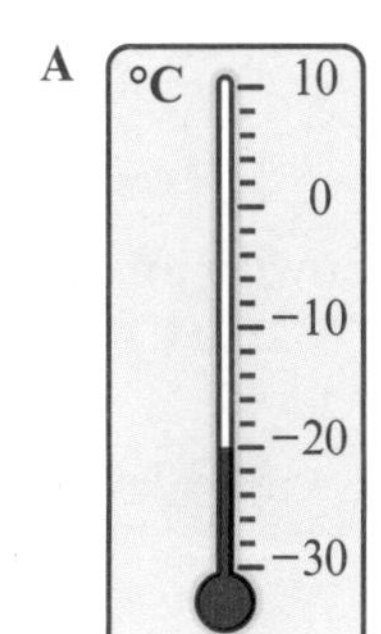

B

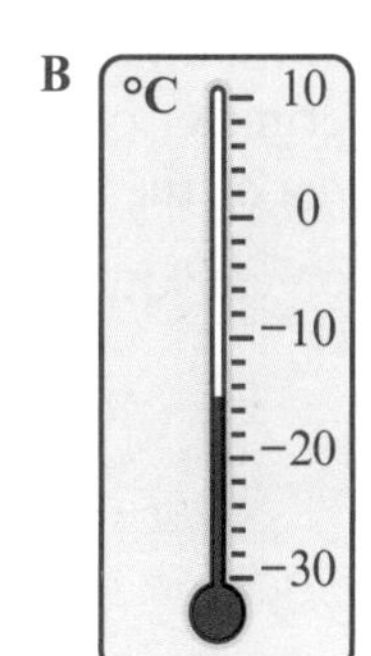

C

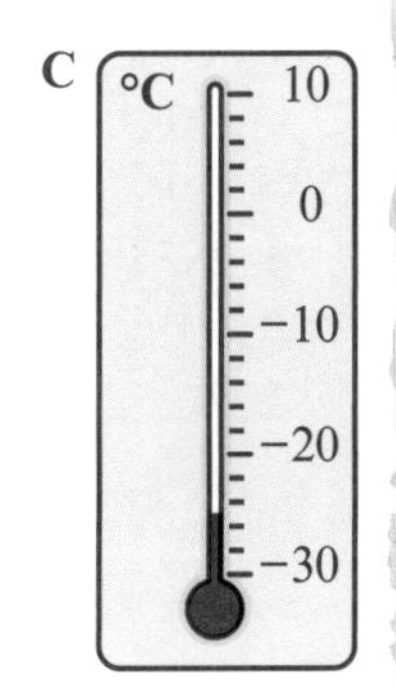

Perimeter and area

Make sure you know how to work out the area and perimeter of simple shapes.

Perimeter

Perimeter is the distance around the edge of a shape. A builder who builds a fence around a playground will need to know the perimeter of the playground.

You can work out the perimeter of a shape by adding up the lengths of the sides.

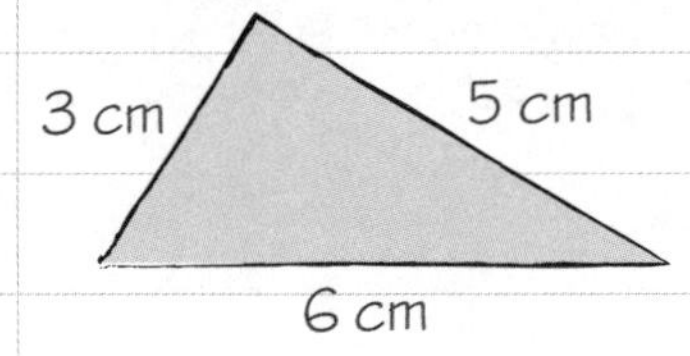

Perimeter = 3 cm + 5 cm + 6 cm
= 14 cm

Area

Area is the amount of space inside a shape. A painter who paints a wall will need to know the area he needs to paint.

You can work out the area of a shape drawn on squared paper by counting the squares.

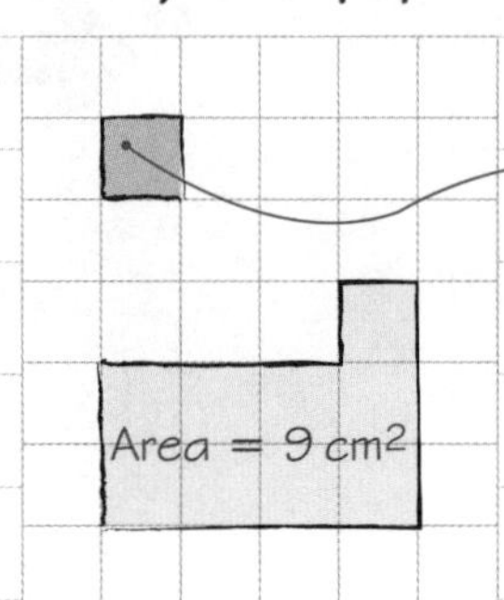

Each square has an area of 1 cm² and sides of 1 cm.

Worked example

This shape is drawn on cm squared paper.

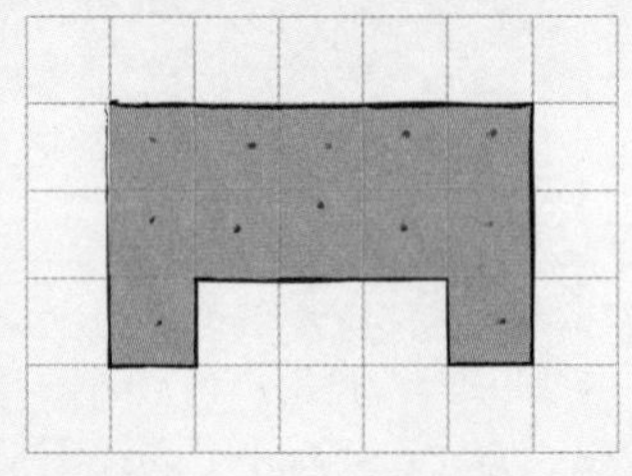

(a) Work out the perimeter of the shape.

18 cm

(b) Work out the area of the shape.

12 cm²

Estimating

You might need to estimate the area of a shape drawn on cm squared paper.

Count 1 cm² for every whole square and for every part square that is more than half full. Do not count squares that are less than half full.

Here there are 10 whole squares.
2 squares are less than half full and 4 squares are more than half full.
A good estimate is 14 cm².

Now try this

1 This shape is drawn on cm squared paper.
  (a) Find the perimeter of the shape.
  (b) Find the area of the shape.

2 Draw a shape on cm squared paper with an area of 20 cm².

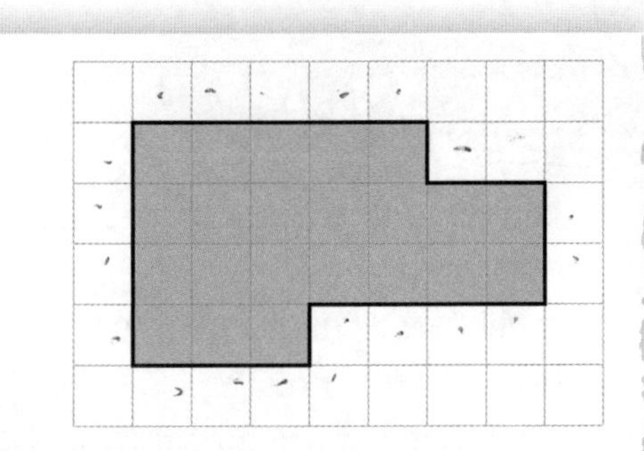

Area of rectangles

You can find the area of a rectangle by counting squares or using a formula.

Area of a rectangle

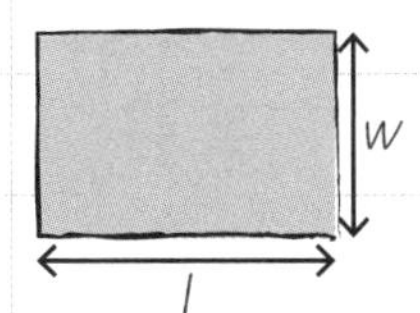

area = length × width
$A = l \times w$

✓ Check the lengths are all in the same units first.

✓ Remember to give units in your answer.

✓ Lengths in cm will give area units of cm^2. Lengths in m will give area units of m^2.

Worked example

1 This diagram shows the floor of a rectangular room. Work out the area of the room.

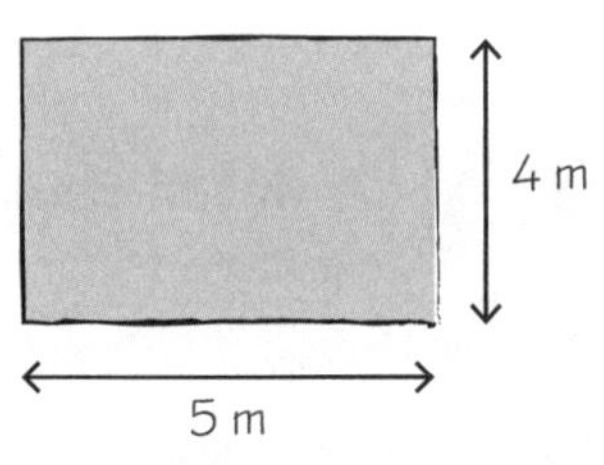

$A = l \times w$

$A = 5 \times 4$

$= 20\ m^2$

The lengths are in metres so the answer is in m^2

Worked example

2 This diagram shows a car park.

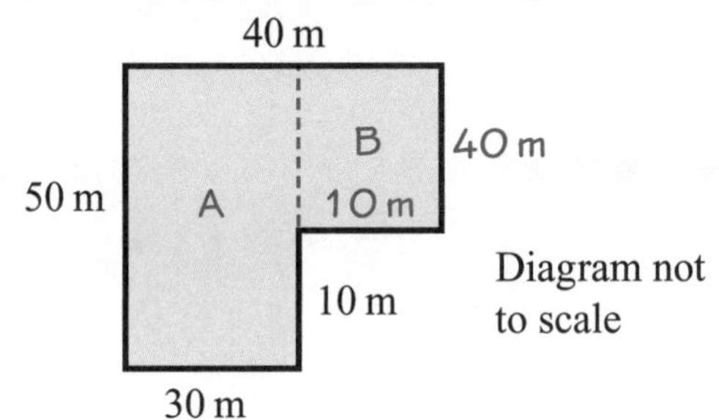

(a) Work out the area of the car park.

area of A = $50 \times 30 = 1500\ m^2$

area of B = $10 \times 40 = 400\ m^2$

total area = $1500 + 400 = 1900\ m^2$

(b) Calculate the perimeter of the car park.

$50 + 40 + 40 + 10 + 10 + 30 = 180\ m$

Golden rule

Work out any missing lengths before calculating the area or perimeter.

Problem solved!

✓ Draw a dotted line to divide the carpark into two rectangles.

✓ Label your two rectangles A and B. You have to use the information in the question to find out the missing lengths.

$40\ m - 30\ m = 10\ m$

$50\ m - 10\ m = 40\ m$

Write these lengths on your diagram.

✓ Find the area of both shapes then add them together to find the total area.

Now try this

1 Find the area of this rectangle.

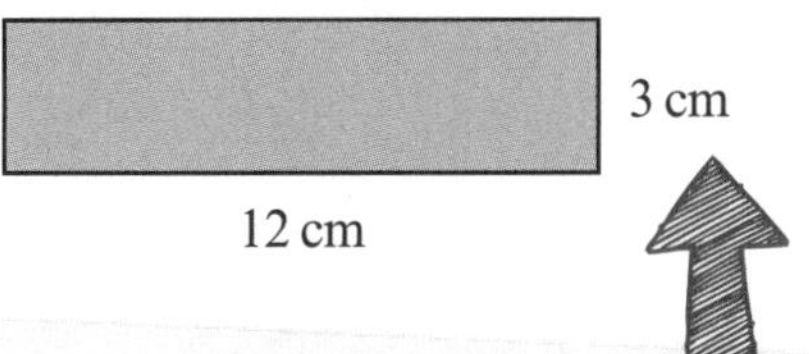

Don't forget to write the units in your answer.

2 A gardener is going to spread some grass seeds on the lawn.

(a) Calculate the area of the grass.

(b) A bag of grass seed covers $4\ m^2$. How many bags of grass seed does the gardener need?

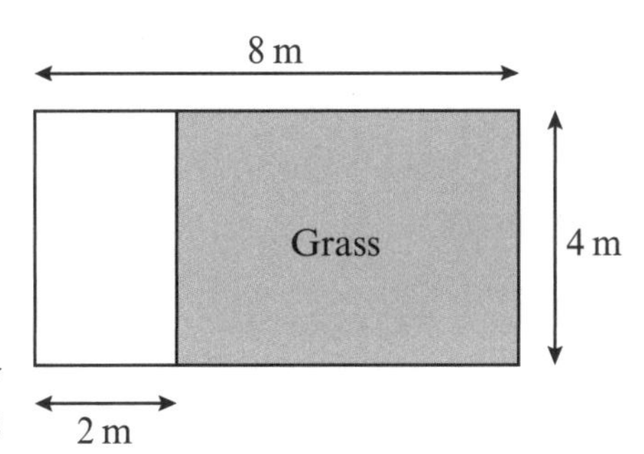

Had a go ☐ Nearly there ☐ Nailed it! ☐

Problem-solving practice

When you are solving problems, you need to:

- read the question
- check your answers
- decide which calculation you are going to use
- make sure you have answered the question asked.

A water cooler has 25 litres of water in it.

A cup holds 150 millilitres of water.

How many cups can be completely filled from the water cooler?

Show your working.

(3 marks)

Capacity page 51

Capacity page 51

Change the units of capacity so that you are working in ml.

You will need to divide to find the number of cups.

TOP TIP

When converting from litres to millilitres, multiply by 1000

Remember, the question asks how many cups can be filled completely, so partly filled cups cannot be counted.

This scale has three identical spheres on it.

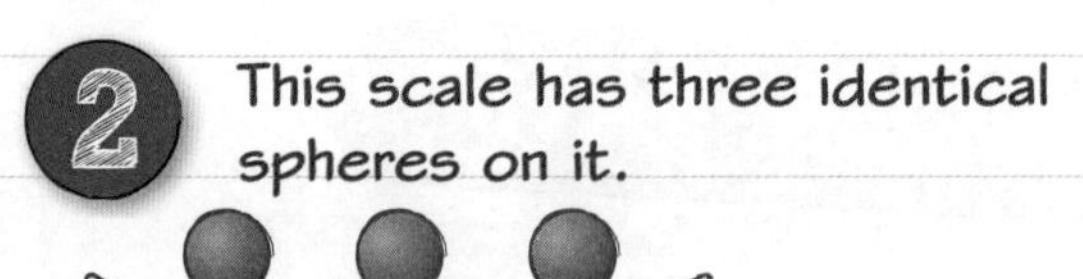

Work out the weight of one sphere.
Give your answer in grams.

(2 marks)

Scales page 46

You can find the weight of the three spheres by reading the scale.

TOP TIP

Make sure you answer the question that has been asked. You need to find the weight of just one sphere, so you have to divide by 3.

A shop sells cartons of juice for 75p each. As a special offer, you can buy three cartons for £1.80

How much money does the special offer save you compared to the full price of three cartons?

(3 marks)

Money page 52

Work out the cost of three cartons without the special offer. Subtract the special offer price from this amount to find the difference.

TOP TIP

Convert the amounts so that they are both in either pounds or pence. Make sure you give the correct units in your answer.

Problem-solving practice

A salesperson needs to visit company offices around the UK.

She starts at the Ashford office and then drives to each of these offices in this order:

Ashford to Horwich

Horwich to Crewe

Crewe to Lymington

Lymington to Ashford.

Ashford			
224	Crewe		
265	42	Horwich	
122	192	230	Lymington

(a) How many miles did she travel altogether? **(2 marks)**

(b) The company pays travel expenses at the rate of 50p per mile. Work out the total travel expenses. **(1 mark)**

Mileage charts page 47

(a) Use the mileage chart to work out the distances for each journey.

Add these up to find the total distance travelled.

(b) Find the total travel expenses by multiplying the number of miles travelled by 50.

To convert from pence to pounds, remember to divide by 100.

TOP TIP

The salesperson makes four different journeys so you should have four different distances to add up.

The map shows the driving distances between different cities.

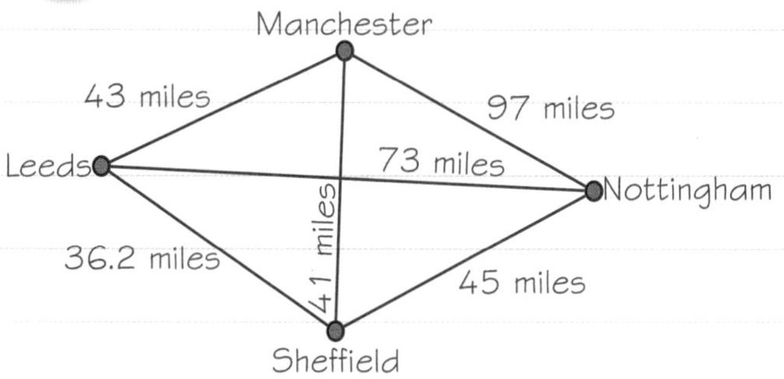

(a) Joseph drives from Manchester to Sheffield and then to Leeds.

How many miles is this? **(1 mark)**

(b) Complete the mileage chart to show the driving distances between the four cities.

Manchester			
	Leeds		
		Nottingham	
			Sheffield

(2 marks)

(c) Joseph leaves his office in Sheffield. He visits all the cities. He then goes back to his office.

Plan Joseph's shortest route. **(3 marks)**

Mileage charts page 47
Routes page 48

(a) Put your finger on Manchester and trace the journey. Add up the total distances.

(b) Look at the distances on the map. The distance from Manchester to Leeds is 43 miles so complete this box in the mileage chart.

(c) Look at the shortest distances on the map. Use the map to help you decide on a possible route. Check by finding the total distance.

TOP TIP

Tick off each journey as you have counted the distance so you can be sure you have covered each one.

Look at all the distances from Manchester and complete this part of the table first.

Now look at all the distances from Leeds.

Had a go ☐ Nearly there ☐ Nailed it! ☐

Problem-solving practice

Marcus is redesigning his garden and wants to add a patio.

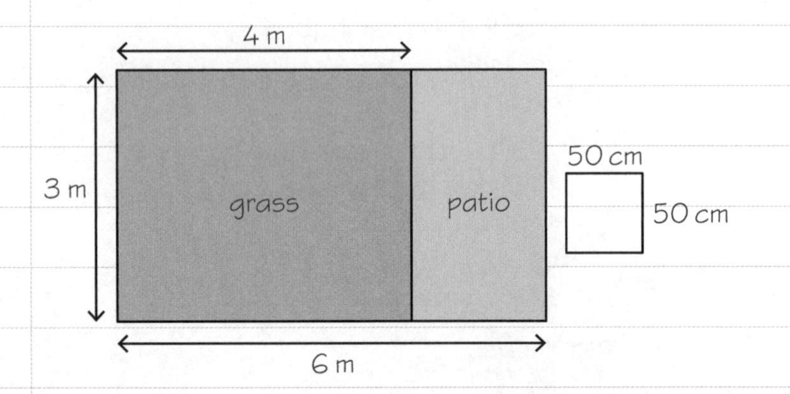

He makes his patio using 50 cm by 50 cm paving slabs.

(a) Paving slabs are sold in packs of five.

How many full packs of paving slabs does Marcus need?

(3 marks)

(b) The cost of a pack of five paving slabs is £6.

How much will it cost Marcus to buy enough slabs to fill the patio?

(1 mark)

Area of rectangles page 55

(a) Convert all lengths to either cm or m.

(b) Marcus may need to buy more slabs than he actually needs as they are sold in packs.

(c) You can work out the total cost by multiplying the number of packs by the cost of each pack.

TOP TIP

In the online test, if you get stuck, you can flag a question to come back to later. Click the review button so that you can check that question again.

A primary school wants to build a fence around its play area. A 2 m gap is needed for a gate.

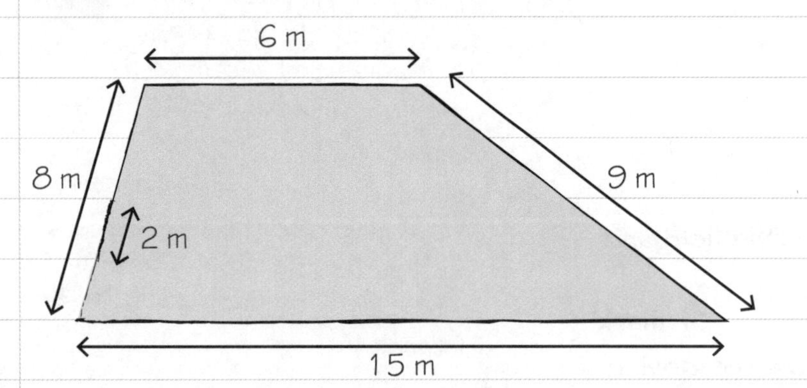

Fencing is sold in 3 m panels. Each panel costs £24.50

Calculate the cost of the fencing needed.

(3 marks)

Perimeter and area page 54

Work out the perimeter of the playground.

A gap is needed for the gate, so don't forget to subtract this.

Each panel is 3 m, so divide the total length by 3 to work out how many panels are needed.

TOP TIP

Make sure you show all of the steps of your working.

Symmetry

You need to be able to spot lines of symmetry and rotational symmetry in shapes. If you tile a room with patterned tiles, you may need to make sure the tiles are in a symmetrical pattern.

Lines of symmetry

A line of symmetry is a mirror line. One half of the shape is a mirror image of the other.

Rotational symmetry

The order of rotational symmetry tells you the number of points in one full turn that a shape looks the same.

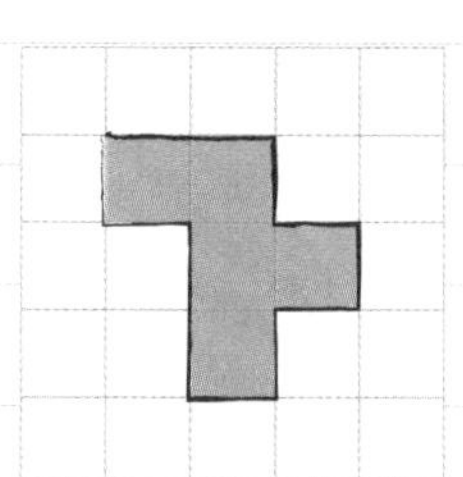

no lines of symmetry

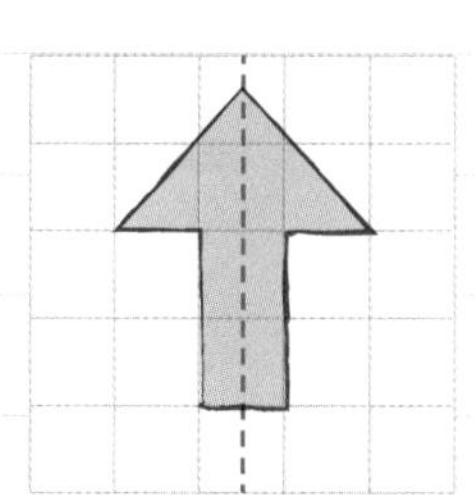

1 line of symmetry

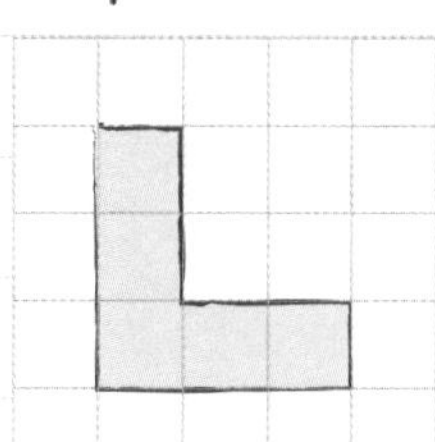

no rotational symmetry

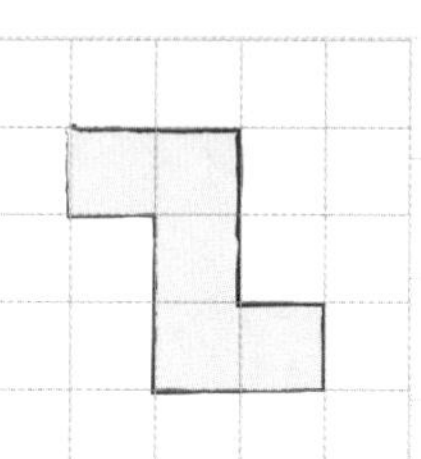

rotational symmetry of order 2

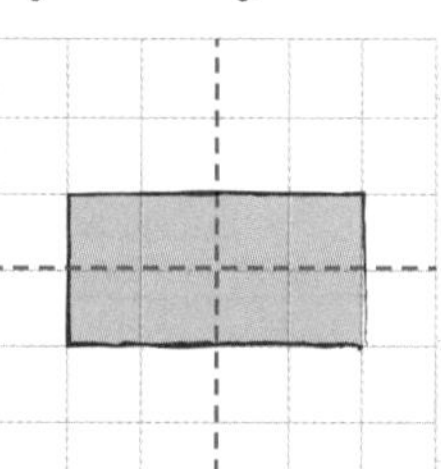

2 lines of symmetry

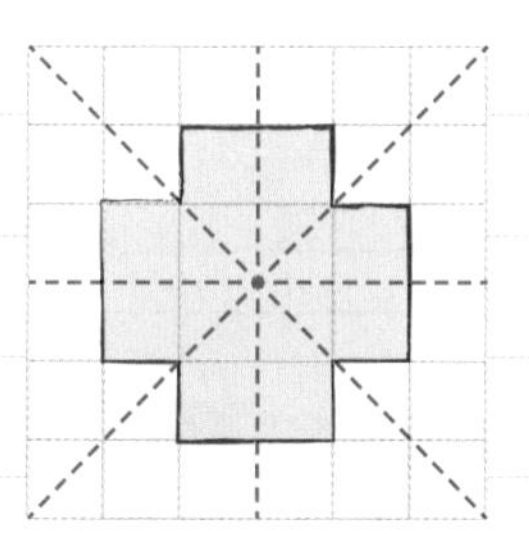

4 lines of symmetry

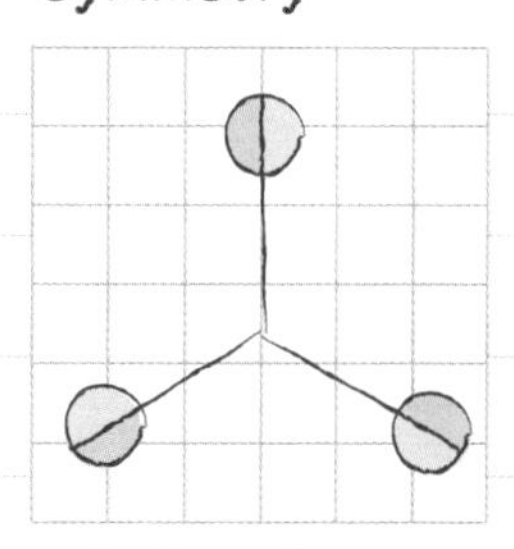

rotational symmetry of order 3

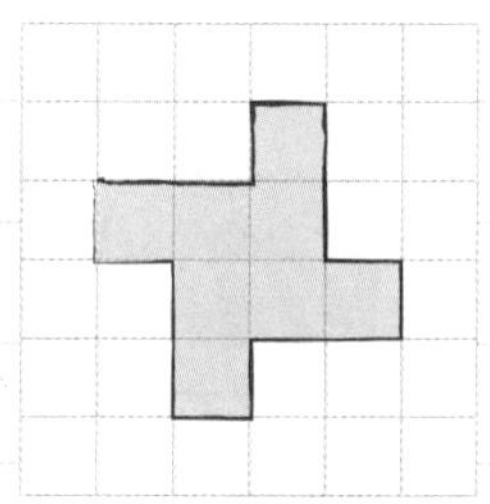

rotational symmetry of order 4

Checking for lines of symmetry

You are allowed to ask for tracing paper in your exam. You can use it to check for lines of symmetry.

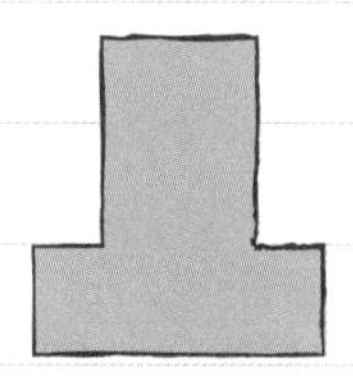

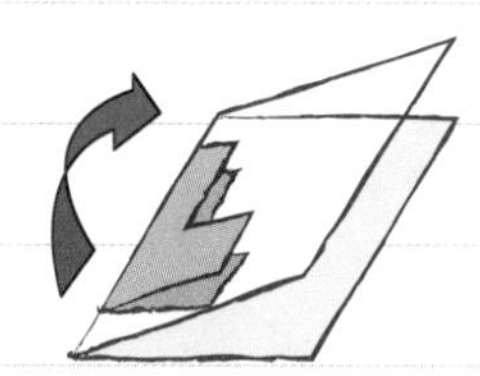

If you fold a tracing of a shape in half along a line of symmetry, the two halves will match up exactly.

Checking for rotational symmetry

You can also use tracing paper to check for rotational symmetry.

Trace the shape. Rotate the tracing paper and see at how many points the shape looks the same.

This shape fits over itself twice: once at 180° and once at 360°. It has rotational symmetry of order 2

Now try this

1 Copy the grid below and shade 5 squares to make a shape with exactly 4 lines of symmetry.

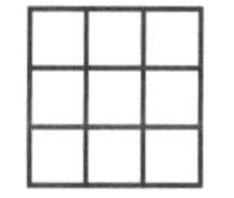

2 Copy the grid below and shade 5 squares to make a shape with rotational symmetry of order 2.

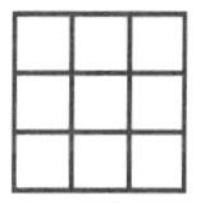

Properties of 2D shapes

You need to be able to recognise 2D shapes.

Naming 2D shapes

Make sure you know the names of these 2D shapes.

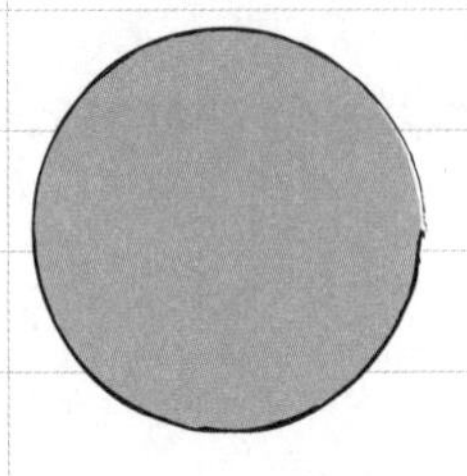
circle

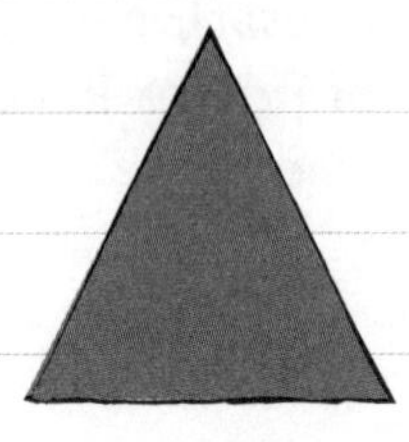
triangle

square

rectangle

Regular shapes

If a shape has sides of all the same length and angles all of the same size, it is **regular**.

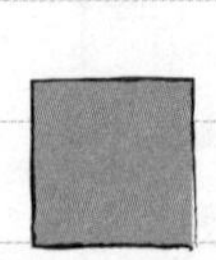
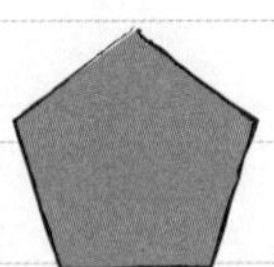
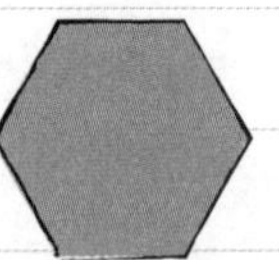
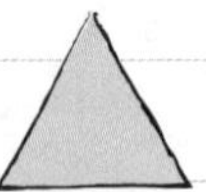

Irregular shapes

If a shape has sides of different lengths and angles of different sizes, it is **irregular**.

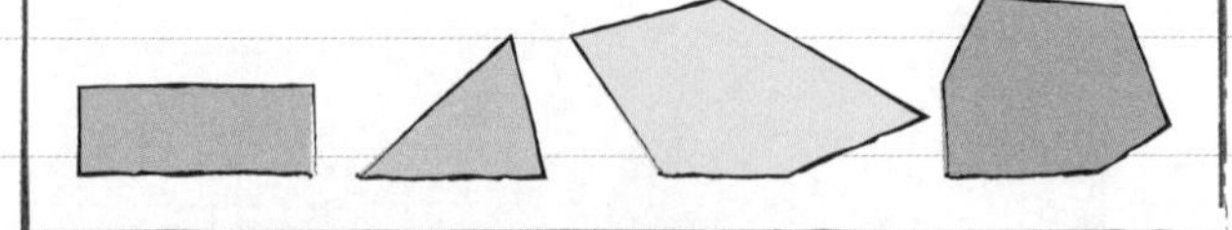

Tiling

Some 2D shapes can fit together side-by-side without any gaps, like square or rectangular bathroom tiles.

You can work out how many tiles you would need to fill a certain area by finding the area of one tile and dividing the whole area by your answer.

Start by converting all of the measurements into the same units. The area is in cm^2 so convert the length in m to cm.

Worked example

Joey wants to tile the splashback behind his sink. He needs to tile an area that is 1.2 m by 40 cm. The tiles he chooses each have an area of $30\,cm^2$. How many tiles does he need to buy to fully tile the splashback?

1.2 m = 120 cm

$120 \times 40 = 4800\,cm^2$

$4800 \div 30 = 160$ tiles

Now try this

1 Name each shape and explain whether it is regular or irregular.

2 Shahina has 360 tiles. They each have a length of 24 cm and a width of 12 cm. She wants to tile a floor that is 4.32 m long by 2.4 m wide. Can Shahina tile the full area without cutting any tiles?

Scale drawings and maps

Scales are used on maps and scale drawings so that you can accurately work out real distances.

Using scales

This is a scale drawing of a plane.

The scale used is 1 cm = 10 m.

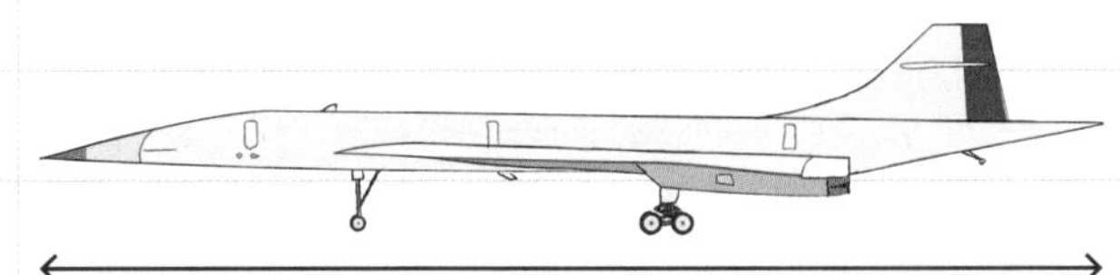

8 cm

You can use the scale to find the real length of the plane.

The length of the plane in the scale drawing is 6 cm.

$8 \times 10 = 80$

The plane is 80 metres long.

Worked example

This diagram shows a scale drawing of an area of Norfolk.

scale: 1 cm = 3 km

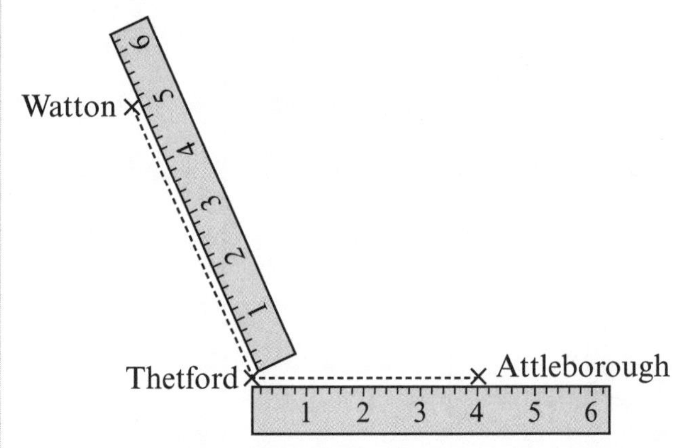

What is the distance between Watton and Thetford?

$3 \times 5 = 15$ km

Read off the distance from Watton to Thetford on the ruler.

scale distance = 5 cm

Then work out the real distance using the scale.

map distance	real distance
1 cm	3 km
5 cm	15 km

(× 5 on both columns)

Now try this

This map shows some of the attractions at a zoo.

(a) Work out the real distance between the coffee shop and the lion enclosure.

(b) Paula said the distance between the giraffe enclosure and the lion enclosure is less than 250 m. Use the map to show whether she is correct or not.

(c) Paula has opened a new children's area at the zoo. She wants to add it to the map. The distance between the giraffe enclosure and the children's area is 400 m. How far will this be on the scale map?

lion enclosure ×

× giraffe enclosure

scale:
1 cm = 50 m

× coffee shop

Had a look ☐ Nearly there ☐ Nailed it! ☐

Using plans

A plan shows the layout of objects in a space as if you are looking directly down from above.

Reading a plan

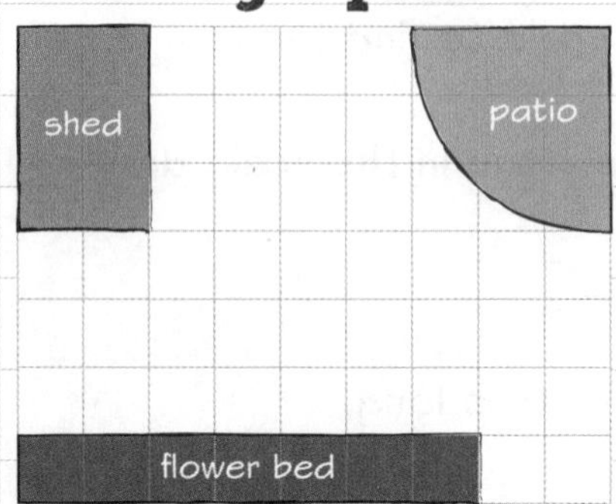

Key:
1 square on the grid is
1 m by 1 m in the garden

This plan shows the layout of objects in a garden.
The key tells you the size of each square on the grid.
The garden is 9 m by 7 m. The shed is 2 m by 3 m.

There are different positions where the bed would fit. Try different places and check that the conditions for where the bed must be placed are met.

Worked example

Jim wants to work out where to put his bed in his new bedroom.

His bedroom is 4 m by 4.5 m.

The bed needs a rectangular space 1.5 m by 2 m.
The bed must be:

- at least 1 m from the window
- at least 1 m from the wardrobe
- at least 1.5 m from the door
- at least 1 m from the shelf.

Jim draws a plan of his bedroom.

Draw the bed on the plan. Remember to use the key.

Key:
1 square = 0.5 m by 0.5 m
——— represents a wall

door
wardrobe
shelf
bed
window

Now try this

Sarah wants to work out where to put a dining table in her dining room.
The table needs a rectangular space with an area of 2.5 m by 1.5 m.
The table must be:

- at least 0.5 m from the window
- at least 1 m from the sofa
- at least 0.5 m from any wall
- at least 0.5 m from the shelves
- at least 1.5 m from the door.

Sarah draws a plan of her dining room.
Copy out the plan onto squared paper and draw the table in a suitable position. Remember to use the key.

Key:
1 square = 0.5 m by 0.5 m
——— represents a wall

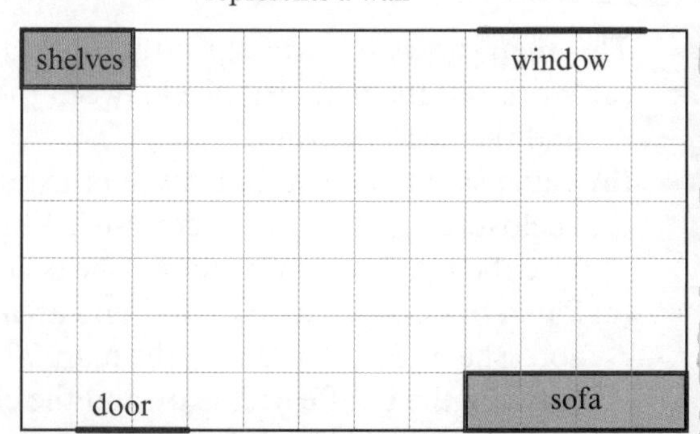

Angles

Types of angle

You need to know the names of the different types of angles.

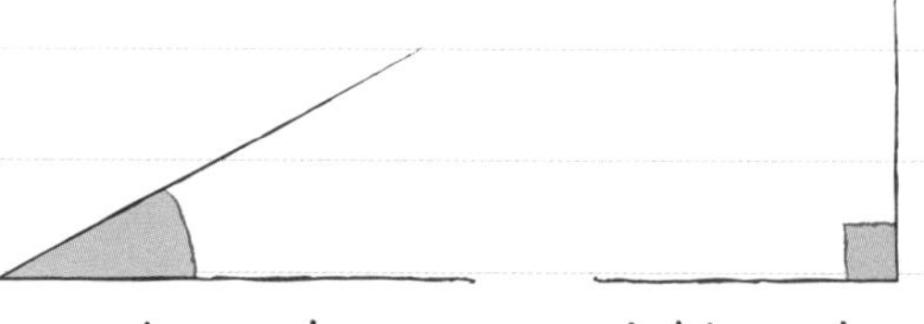

acute angle less than 90° | right angle 90° | obtuse angle between 90° and 180° | reflex angle more than 180°

Measuring angles

A protractor measures angles in degrees. Use the scale on the protractor that starts with 0 on one of the lines of the angle.

Here, use the outside scale.

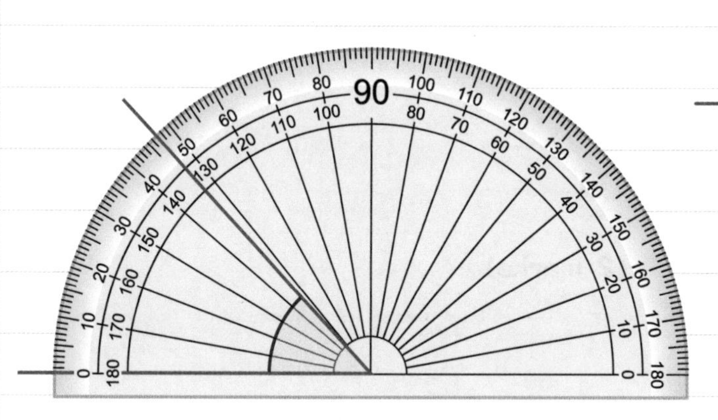

Place the centre of the protractor on the point of the angle. Line up the zero line with one line of the angle. Read the size of the angle off the scale. This angle is 47°.

Estimate the size of an angle before measuring it. This lets you check that your answer is sensible.

Here, use the inside scale.

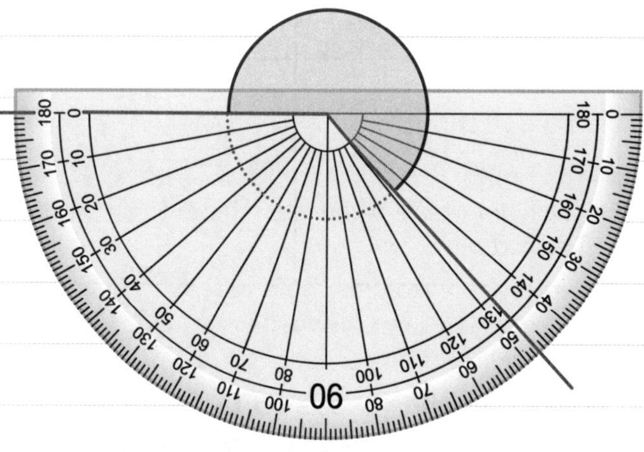

To measure an angle bigger than 180°, measure the missing part of a 360° angle (shown by the dotted line), then subtract the answer from 360°.

$360 - 133 = 227°$

The marked angle is 227°.

Now try this

Name the types of angles shown in these diagrams and write an estimate of the size of each angle. Then measure each angle accurately with a protractor.

(a)

(b)

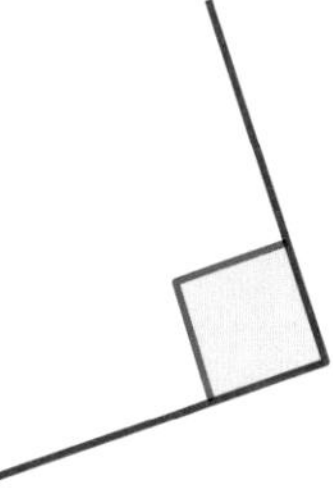

(c)

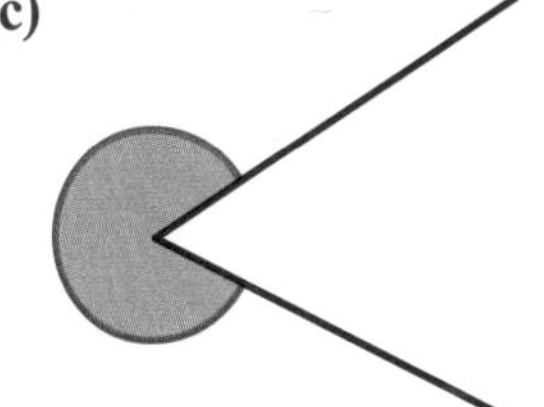

Had a go ☐ Nearly there ☐ Nailed it! ☐

Problem-solving practice

When you are solving problems, you need to:

- read the question
- check your answers
- decide which calculation you are going to use
- make sure you have answered the question asked.

1 Ms Khan is designing a tiling pattern using four blue and white tiles. The pattern must have two lines of symmetry. Complete the diagram by shading the tiles to show a suitable design.

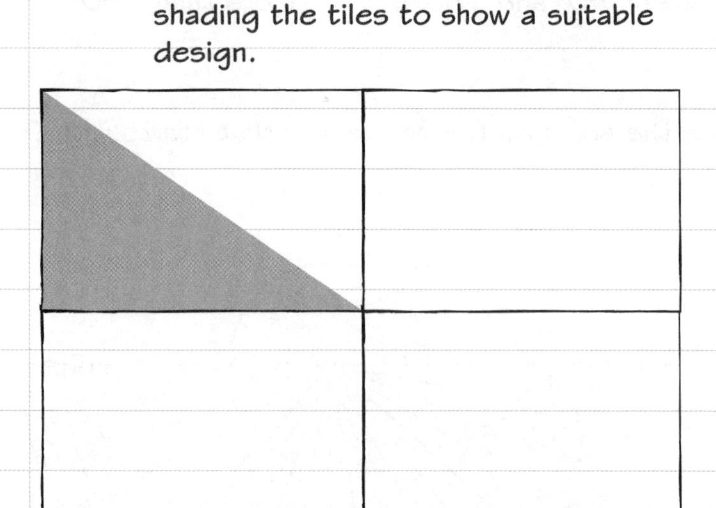

(2 marks)

Symmetry page 60

The shape needs to have exactly two lines of symmetry. There are lots of different ways to solve this question.

TOP TIP

Check that your answer has exactly two lines of symmetry.

2 Gunther needs to buy enough paving stones to pave a market square. He measures the market square and draws this plan.

42 m

25 m

A consultant tells Gunther that she can provide enough paving stones to cover $1000 m^2$. Will the paving stones cover a large enough area?

(2 marks)

Properties of 2D shapes page 61

Work out the area of the town square first. Then decide whether there will be enough paving stones to cover it.

TOP TIP

Make sure you answer the question asked. You need to write a statement to say whether or not there are enough paving stones.

Problem-solving practice

3 The map shows the areas in a park. The scale used is 1 cm = 40 m.

lake

car park

children's play area

cafe

(a) Work out the real distance between the car park and the lake. **(1 mark)**

(b) Sandeep walked from the car park to the cafe and then to the children's play area. How far did he walk? **(2 marks)**

Scale drawings and maps page 61

The scale is important.

Measure the distance between the car park and the cafe with a ruler.

Use the scale to work out the real distance.

TOP TIP

Check your answer is sensible.

4 Ruth has bought some shelves for her living room. The room is 3.5 m by 4.5 m.

The shelves need a rectangular space 1 m by 0.5 m. They must be:

- on a wall
- at least 1 m from the sofa
- at least 0.5 m from the window
- at least 1 m from the door
- at least 1 m from the TV.

Ruth draws a plan of her living room.

Key:

1 square = 0.5 m by 0.5 m

——— represents a wall

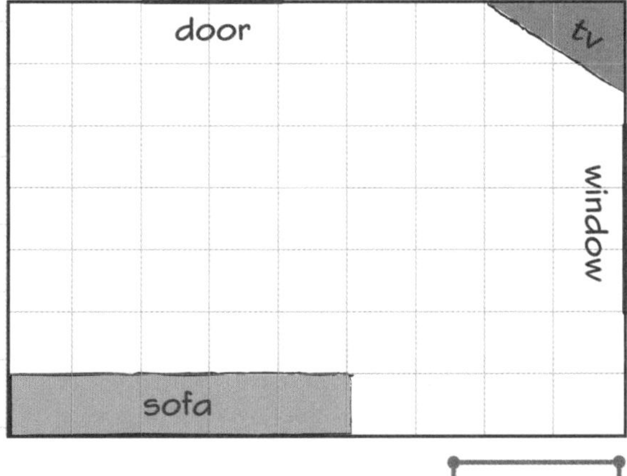

shelves

Copy the plan and show where she can put the shelves. Remember to use the key.

(2 marks)

Using plans page 62

There are different positions where you can put the shelves. Try different positions and check that the conditions for where the shelves can be placed are met.

TOP TIP

On the online test, the shelves would be an icon.

You need to make sure the icon is the correct size by dragging the dots.

You can then drag the icon around the grid.

Tables

You need to understand data displayed in tables so that you can solve problems.

Understanding tables

Speed	Stopping distance
20 miles per hour	12 metres
30 miles per hour	23 metres
40 miles per hour	36 metres

This table shows data about car stopping distances at different speeds. At 30 miles per hour, the stopping distance is 23 metres.

Worked example

1 This table shows information about the cost of booking a hotel room during peak and off peak seasons.

	Off-peak	Peak
Standard double	£70	£105
Standard twin	£72	£110
Deluxe double	£120	£180
Deluxe twin	£135	£200

How much will it cost to book a standard double room for 3 nights during off-peak season?

$3 \times 70 = £210$

Look down the first column until you find the standard double room.

Read across to the amount under the 'off-peak' heading.

Multiply this amount by 3 to find the price for three nights.

To find the difference in mileage, you need to find the mileage for each car on the table and then subtract the smaller from the larger.

Worked example

2 This table shows information about two second hand cars for sale.

	Car A	Car B
Engine size	1300 cc	1600 cc
Mileage	42 000	30 000
Cost	£5,000	£7,000

(a) What is the cost of car A?

£5,000

(b) What is the engine size of car B?

1,600 cc

(c) Work out the difference in mileage between the two cars.

$42000 - 30000 = 12000$ miles

Now try this

The table shows information about the hours and pay of two employees one week.

	Employee 1	Employee 2
Normal hours worked	25	24
Overtime hours worked	5	8
Pay per hour for normal hours	£12	£10
Pay per hour for overtime hours	£18	£20

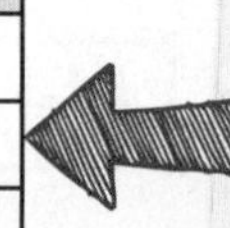

Find the necessary information in the table then calculate the total amount each employee earned.

Who earned more money over the week?

Tally charts and frequency tables

A tally chart is used to record information collected from an experiment or survey.

A frequency table shows the number of times something occurs.

This tally and frequency table shows the number of drinks sold in a cafe one morning.

A tally is useful for collecting information. You can represent each drink sold with a line.

Number of drinks sold	Tally	Frequency
tea	|||	3
coffee	~~||||~~ ||	7
fruit juice	~~||||~~ ~~||||~~ ||||	14
water	~~||||~~ |	6

To work out the frequency, add up the tallies. The frequency tells you the number of drinks sold. The number of teas sold is 3.

You represent 5 drinks by writing 4 lines with a fifth line through them.

Worked example

1 A hairdresser collects information about the services her customers book one week.
The information is shown in this tally chart.

Service	Tally	Frequency
cut	~~||||~~ ~~||||~~	10
blow dry	~~||||~~ ~~||||~~ ||	12
highlights	~~||||~~ ||||	9
colour	|||	3

(a) Complete the table.

(b) Which was the most popular service?

blow dry

(c) How many customers booked highlights?

9

Worked example

2 A florist sold 100 flowers in one day.
He recorded the type of flowers.
The frequency table shows this information but some information is missing.

Flowers	Frequency
roses	21
tulips	34
lilies	
carnations	12

(a) Find the missing value.

$21 + 34 + 12 = 67$

$100 - 67 = 33$

(b) How many more roses than carnations were sold?

$21 - 12 = 9$

Now try this

A museum shop recorded the items bought during one day.

(a) Copy and complete the tally and frequency table.

(b) How many items were bought from the shop?

Item	Tally	Frequency
postcards	~~||||~~ ~~||||~~ ~~||||~~ ~~||||~~ ||	
key rings	~~||||~~ ~~||||~~	
book		13
toy	~~||||~~ ~~||||~~ |	
poster	~~||||~~ |||	

Had a look ☐ Nearly there ☐ Nailed it! ☐

Data collection sheets

A data collection sheet is a table designed to record specific data. When you design a data collection sheet, think about how you can make it quick and easy to use.

Designing a data collection sheet

Here is a data collection sheet to record what mode of transport four different members of staff use to get to work.

Name	Walk	Train	Bus	Cycle	Other

Choose sensible headings for your columns. Include all the possible choices.

Leave space in the first column for four people's names.

You just need to tick one column for each person.

You can add a column labelled 'Other' to collect responses that don't fit into any of the other categories.

Worked example

Katia is the manager of a cafe. She wants to record how many cups of tea and how many cups of coffee are sold each hour from 9 a.m. to 1 p.m.

Time slot	Tally for tea	Tally for coffee
9 a.m. to 10 a.m.		
10 a.m. to 11 a.m.		
11 a.m. to 12 a.m.		
12 a.m. to 1 p.m.		

Think about the different time slots that you want to record data for. Each time slot should have its own row.

Include a column for tea and a column for coffee so that Katia can use a tally to count how many of each type of drink are sold.

Problem solved!

In the test, make sure your data collection sheet would be **efficient** to use. Think about whether you could use a tally or ticks to record data rather than a written list.

Now try this

1 A car manufacturer wants to collect information about the colour of cars people want. She is going to ask 30 customers to choose between red, blue, black, white, silver and other. Design a data collection sheet to collect this information.

2 A manager wants to find out about the level of motivation among his six members of staff.
For each member of staff he wants to record:
- the person's name
- the person's job title
- whether they are 'happy', 'OK' or 'dissatisfied' at work.

Design a data collection sheet to collect this information.

Reading bar charts

You can use a bar chart to represent data given in a tally chart or frequency table.
This bar chart shows the number of houses sold by an estate agent each month.

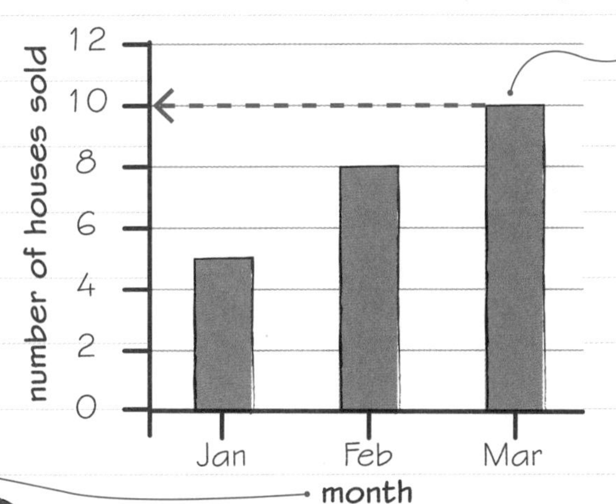

The height of the bar shows you the frequency, or the number of houses that were sold.

Both axes are labelled.

Worked example

This bar chart shows the number of mobile phones sold by Rosie in one week.

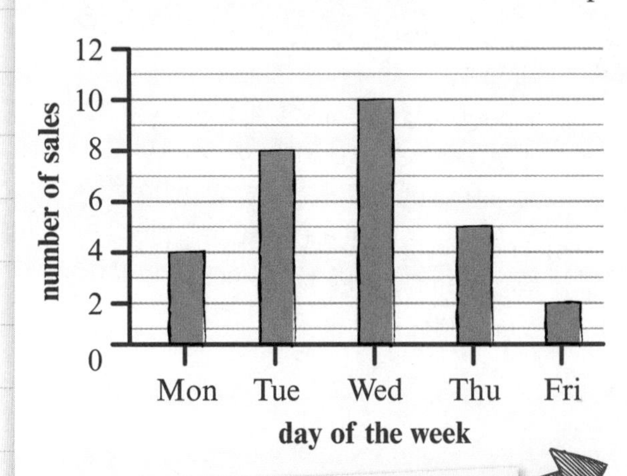

Wednesday was her best day and Friday was her worst day.

(a) How many phones did Rosie sell on Tuesday?

8

(b) On which day did Rosie sell the greatest number of phones?

Wednesday

(c) What is the difference in the number of phones she sold on her best day and worst day?

$10 - 2 = 8$

(d) How many phones did Rosie sell in total during the week?

$4 + 8 + 10 + 5 + 2 = 29$

Now try this

This bar chart shows the number of flowers sold by a florist in one day.

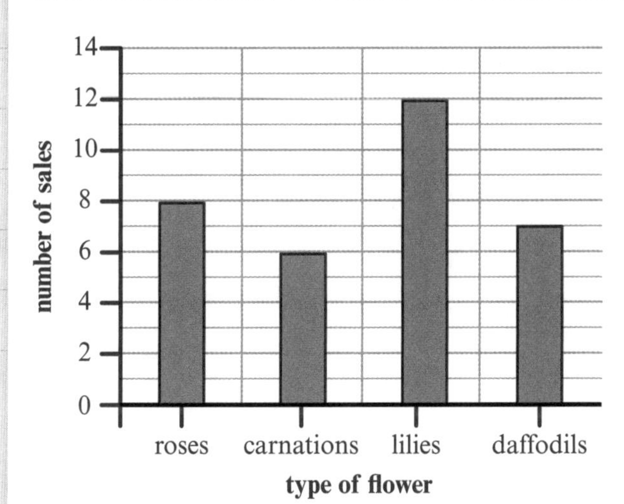

(a) Write a statement about the number of flowers sold.
(b) How many flowers were sold altogether?

If you are asked to make a statement about a chart, there are lots of possible answers you could give. For example, you could say which type of flower the florist sold most of or which type of flower the florist sold least of.

Had a look ☐ Nearly there ☐ Nailed it! ☐

Reading pictograms

You can use a pictogram to represent data from a tally chart or frequency table.

Understanding pictograms

A pictogram uses pictures to represent different amounts. The key tells you how many items each picture represents.

Worked example

This pictogram shows the number of films Adam watched in June, July, August and September.

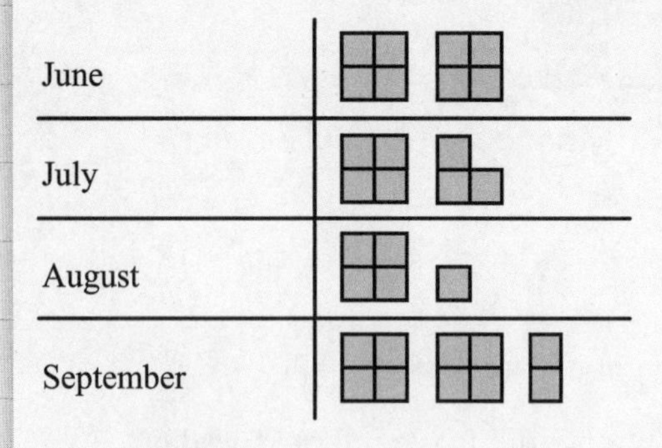

(a) Write down the number of films Adam watched in June.

8

(b) Write down the number of films Adam watched in August.

5

(c) In which month did Adam watch the most films?

September

Use the key to work out what each picture represents.

= 2 films = 1 film

(a) There are 2 blocks of 4 squares in June. This represents $2 \times 4 = 8$ films.

(b) There is a block of 4 squares and 1 extra square in August. This represents $4 + 1 = 5$ films.

(c) The month with the most blocks is September, so that must be when he watched the most films.

Now try this

The pictogram shows the number of people who viewed a house over three days.

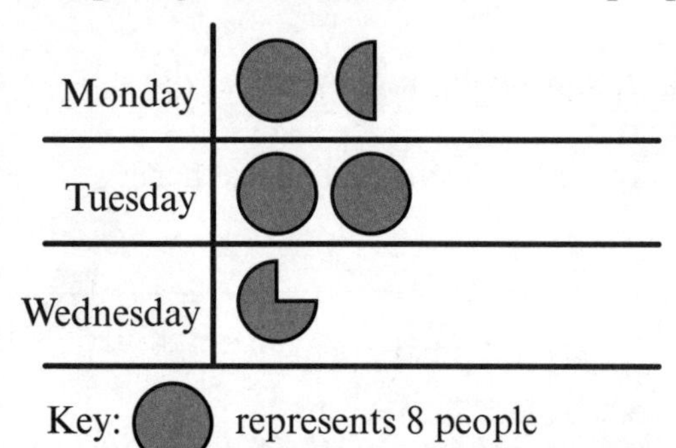

(a) How many people viewed the house on Wednesday?

(b) How many more people viewed the house on Tuesday than Monday?

(c) How many people viewed the house in total over the three days?

Reading pie charts

A pie chart is a circle divided into slices.
Each slice represents a proportion of the whole.
This pie chart represents the colour of buckets in a toy shop.

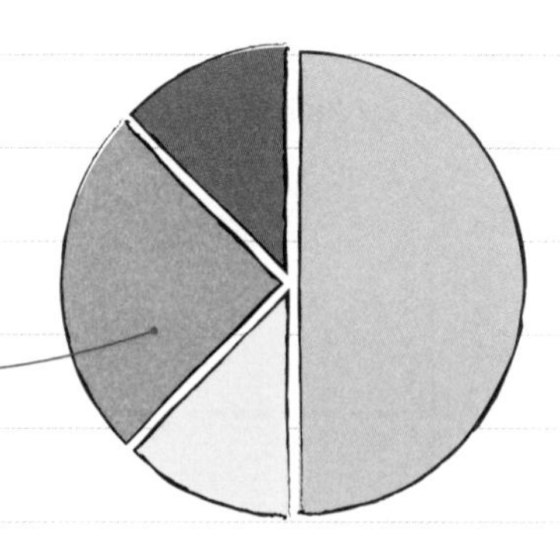

Worked example

The pie charts show information about the number of items recycled by two families last month.

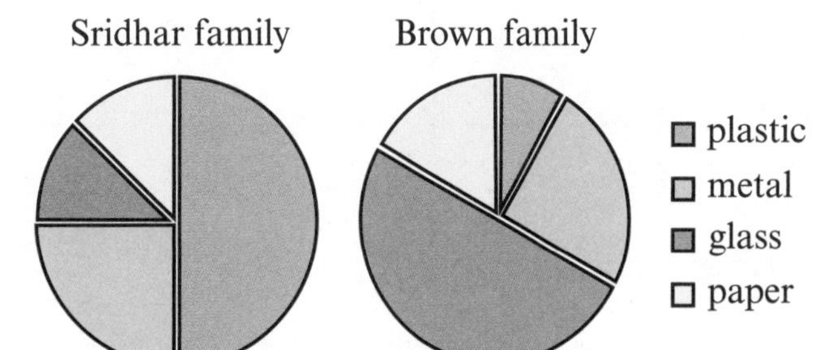

(a) What item did the Sridhar family recycle most of?

plastic

(b) What fraction of the total amount of items recycled by the Sridhar family were metal?

$\frac{1}{4}$

(c) The Brown family recycled 70 glass items. How many items did they recycle in total?

140

(d) 'The Brown family recycled more glass than the Sridhar family'. Is this statement true? Explain your reasons.

You cannot tell because you don't know how many items the Sridhar family recycled in total. The pie chart just shows the proportions.

(a) Look at the largest sector in the pie chart.
(b) Look at the sector that represents metal. What fraction of the pie chart is this?
(c) $\frac{1}{2}$ of the pie chart is glass items.
If $\frac{1}{2}$ is 70 items, the whole is $2 \times 70 = 140$

Now try this

The pie charts show the services customers had at two hair salons in one week.

Customers had either a cut, a blow dry, colour or highlights.

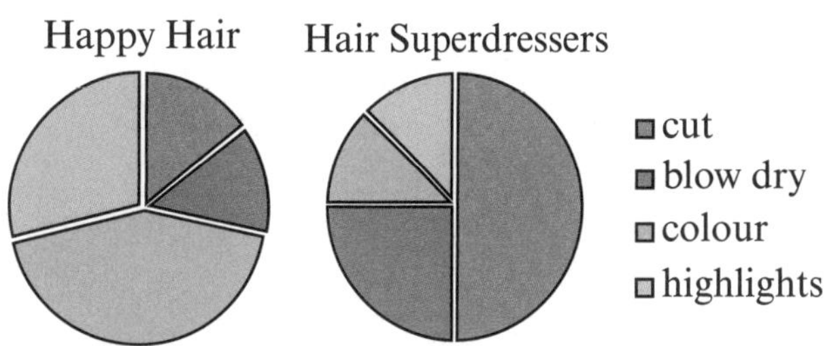

(a) 20 people ordered a cut at Hair Superdressers. How many customers did they have altogether?

(b) Can you tell which hair salon had the most highlights customers? Explain your answer.

Had a look ☐ Nearly there ☐ Nailed it! ☐

Reading line graphs

Line graphs show the relationship between two measurements.

This line graph shows how the average UK income has changed over time.

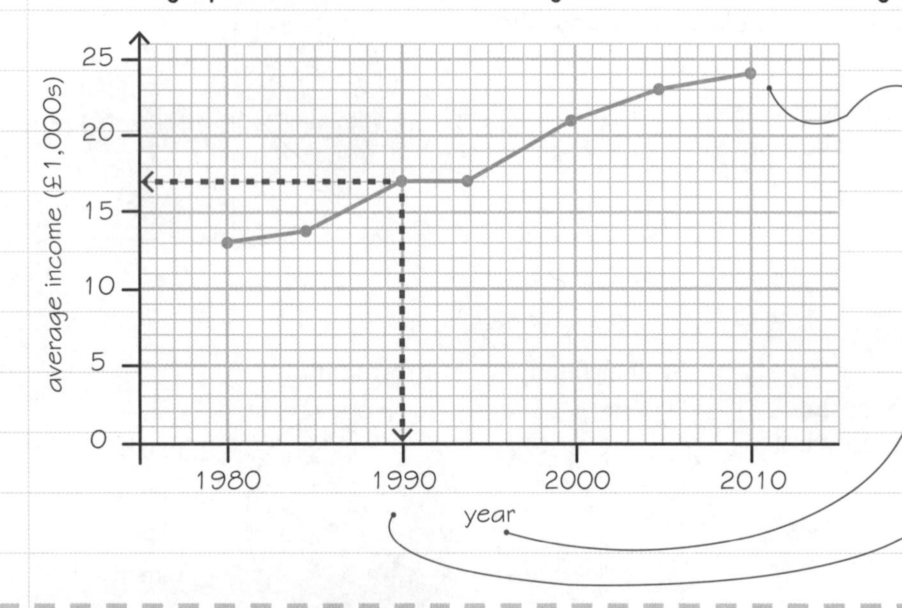

Worked example

Sophie's company pays her 50p for each mile she travels. The graph can be used to work out how much her company pays her for travel.

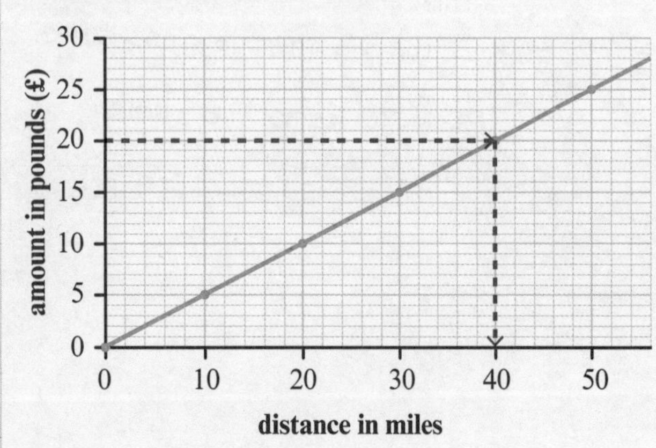

Sophie's company paid her £20
Work out the distance Sophie travelled.

40 miles

Problem solved!

- ✓ Read across from £20 on the y-axis until you get to the line.
- ✓ Move down from the line to the horizontal axis.
- ✓ Read the value from the horizontal axis. This is your answer.

Now try this

Look at the graph in the worked example above. Sophie travels 50 miles. How much will her company pay her?

Planning a graph or chart

You may be asked to draw a suitable graph or chart to show data you are given. You will need to choose which graph to draw and think about the scale, the axes, and the labels and titles.

Scale

- ✓ The y-axis should start at 0 and go up to at least the largest value.
- ✓ Decide how much to go up by each time. It is usually easiest to go up in multiples of 1, 2, 5, 10, 50 or 100.
- ✓ You don't need to use all of the space but try to make the graph as clear as possible.

For a reminder about scales, see page 46.

Axes

- ✓ The data that you measure (such as height or temperature) goes on the **y-axis**.
- ✓ Time (such as hours or years) goes on the **x-axis**.

Labels and titles

- ✓ Label both axes to explain what data they show. You can put units in brackets next to the title rather than on the scale.
- ✓ Give the graph a title to explain what it shows.

Choosing which type of graph or chart to draw

Use a **line graph** when you want to look at the relationship between two sets of data, such as how something changes over time.

Use a **bar chart** when you want to compare data for different categories, such as average rainfall in different cities.

Tips for drawing graphs

- ✓ Draw your graph on graph paper to make it easier to read.
- ✓ Use a pencil to plot your points so that you can easily correct any mistakes.
- ✓ Use a ruler to draw the straight lines.

Worked example

An estate agent collects information about total sales during the last four years.

Year	2012	2013	2014	2015
Sales (£)	200 000	150 000	120 000	150 000

(a) Which data will be on the x-axis? Which data will be on the y-axis?

Year will go on the x-axis as it is time data. Sales will go on the y-axis.

(b) His graph paper has 15 horizontal lines. What would be a suitable scale for the y-axis?

0 to 200 000 in jumps of 20 000

(c) What type of graph should the estate agent draw?

line graph or bar chart

Sometimes, a bar chart or a line graph would both be suitable. Choose the one you prefer to draw.

Now try this

Nawal recorded the height of a plant over four months.

Month	1	2	3	4
Height (cm)	10	12	16	24

(a) Which data will she plot on the x-axis? Which data will be on the y-axis?
(b) Her graph paper has 15 horizontal lines. What would be a suitable scale for the y-axis?
(c) What type of graph should Nawal draw?

Drawing bar charts

You need to be able to draw bar charts.

Bar chart features

- The bars are all the same width.
- There is a gap between the bars.
- Both axes have labels.
- The bars can be drawn horizontally or vertically.
- The height (or length) of each bar represents the frequency.

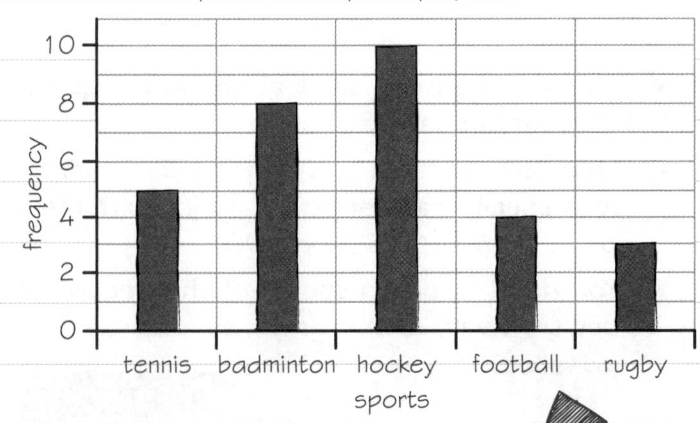

For a reminder about reading bar charts, see page 69.

This bar chart has been correctly drawn. Make sure you are familiar with all of its features so that you can draw your own.

Worked example

This table shows the number of holiday bookings received in one week.

Day of the week	Number of bookings
Monday	12
Tuesday	6
Wednesday	15
Thursday	20
Friday	10

Draw a bar chart to show this data.

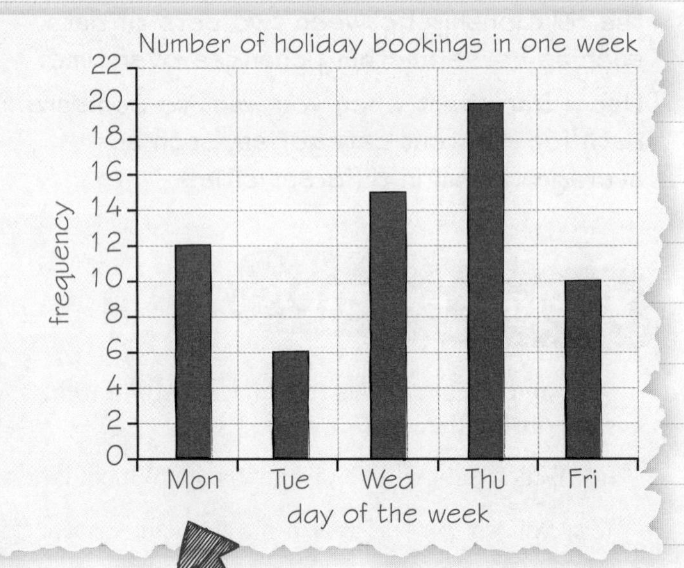

The biggest number in the table is 20, so the scale must go up to at least 20.

Space the numbers on the scale equally. Label the axes.

Now try this

Think about what multiple your scale will go up in.

This table shows the number of job applications Alison sent in one week.

Day of the week	Mon	Tue	Wed	Thu	Fri
Number of applications	6	10	12	13	10

Draw a bar chart to show this data.

Drawing pictograms

You need to be able to draw pictograms.

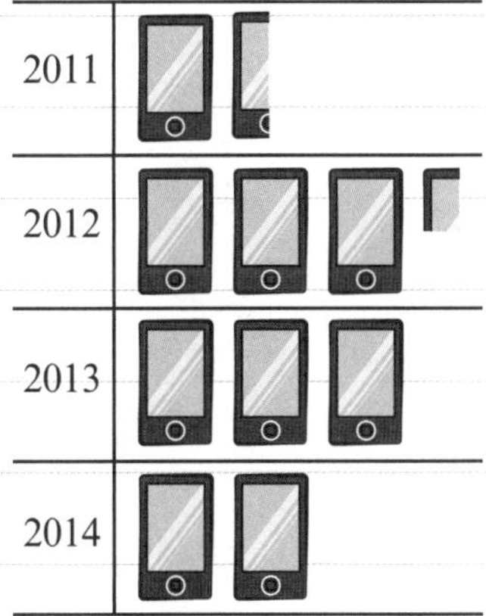

Pictogram symbols

This pictogram shows the number of mobile phones sold by a shop in four years.

A symbol showing a quarter of a phone represents a quarter of 100 = 25 mobile phones.

The pictogram must have a key which tells you how many items are represented by each symbol.

For a reminder about reading pictograms see page 70.

Worked example

This table shows the number of emails Pamela received at work each month.

Month	Number of emails
January	40
February	80
March	45
April	30

Complete the pictogram to show this information.

The key tells you that each envelope represents 20 emails. This means that a quarter of an envelope must represent 5 emails.

January	✉ ✉
February	✉ ✉ ✉ ✉
March	✉ ✉ ◸
April	✉ ◧

Key: ✉ represents 20 emails

Now try this

This pictogram gives information about the number of goals scored in a local football league over 3 weeks.

(a) The number of goals scored in the first week is 12. Draw a key.

(b) How many goals were scored in the third week?

(c) 8 goals were scored in the fourth week.

6 goals were scored in the fifth week.

Copy the pictogram and complete it with this information.

first week	⚽ ⚽ ⚽
second week	⚽ ⚽ ⚽ ⚽
third week	⚽ ◖
fourth week	
fifth week	

Had a look ☐ Nearly there ☐ Nailed it! ☐

Drawing line graphs

You need to be able to draw line graphs.

Worked example

A plumber charges £20 for each hour he works.

Time (hours)	Charge (£)
0	0
1	20
2	40
3	60
4	80
5	100
6	120

The table shows his charges.
Plot his charges on a line graph.

You need to work out the scales for the axes.
The horizontal axis must go from 0 to 6.
The vertical axis must go from 0 to 120.
Remember to label your axes and add a title to your chart.

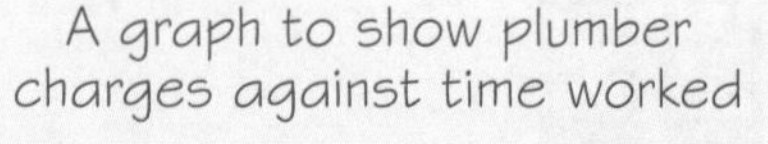

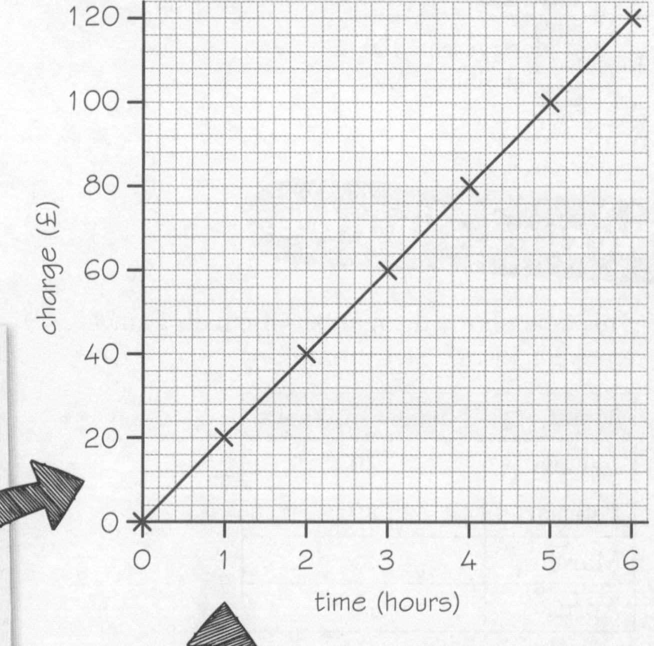

Plot each point.
For 2 hours' work, the plumber charged £40
Put your finger on 2 on the horizontal axis and move up.
Put another finger at 40 on the vertical axis and move across.
Where your two fingers meet, mark a cross.
When you have plotted all the points, join them up with a straight line.

When you have plotted all the points, join them up with a straight line.
Remember to label your axes and add a title to your chart.

Now try this

This table shows some conversions from pounds (£) to dollars ($).
Draw a line graph to show this information.

Pounds (£)	Dollars ($)
0	0
10	15
20	30
30	45
40	60
50	75
60	90

It doesn't matter which currency you put on which axis.

Mean

An average is a value that represents a set of data. The mean is one type of average.

Finding the mean

To find the mean:

add up all the values

divide by the number of values.

Worked example

1 Michael recorded the amount he spent each day on his lunch for five days.

Day	Amount spent
Monday	£4.00
Tuesday	£3.00
Wednesday	£2.50
Thursday	£3.50
Friday	£3.00

Calculate the mean amount he spent each day.

$4 + 3 + 2.50 + 3.50 + 3 = 16$

$16 \div 5 = 3.2$

mean = £3.20

The mean might not be one of the values you are given. It could be a decimal number.
Make sure you answer the question. You are asked to find a mean amount of money, so give your answer in pounds with two decimal places.

Worked example

2 A beauty technician sees 10 clients in 5 hours.

Her target is to see each client for an average of 30 minutes.

Has she met her target?

1 hour = 60 minutes

5×60 minutes = 300 minutes

300 minutes $\div$ 10 = 30 minutes

Yes, she has met her target.

Remember to answer the question asked. The question asks whether she has met her target so you need to write a statement to say whether she has or hasn't.

Now try this

A mortgage adviser recorded the number of customers she had over four weeks.

Week	Number of customers
1	45
2	42
3	31
4	18

(a) Work out her average number of customers.
(b) Her target average number of customers per week is 40. Has she met her target?

Range

You may be asked to work out the range for a set of data. The range is the difference between the largest and smallest values in the data.

Range

To calculate the range, work out:

range = largest value – smallest value

Look for the largest and smallest number of goals scored.

Worked example

1 Here are the numbers of goals scored by a footballer in his last 5 matches.

1 3 2 5 2

Work out the range of goals scored.

Range = 5 − 1
= 4

Worked example

2 The frequency table shows the number of books read in the last month by 25 students.

Number of books	Frequency
1	4
2	7
3	5
4	4
5	5

Work out the range of the number of books read by the students.

5 − 1 = 4

The range is the largest number of books minus the smallest number of books.
A common mistake is to subtract the smallest frequency from the largest frequency.

Now try this

The bar chart shows the number of miles a salesperson travelled to meet customers each month.
Work out the range of the number of miles travelled.

Use the bar chart to find the largest and smallest number of miles travelled. Then find the difference between them.

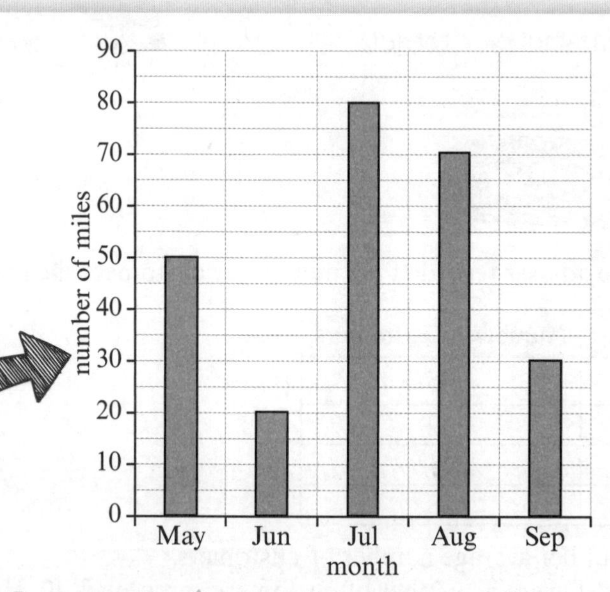

Making a decision

You will need to be able to make decisions based on data. You can use averages and graphs to help you come to your decision.

Worked example

Use the bar chart to decide in which month the family would most like to travel. Use the table to work out when they can afford to do so.

A family of two adults and one child wants to go on holiday to a resort in Spain. They want to go during the warmest month.

The bar chart shows the average temperature in Spain.

The table shows the cost per person per week for different departure dates.

Date of departure	Adult price	Child price
1 June – 1 July	£440	£150
1 July – 21 July	£520	£180
22 July – 12 August	£540	£200
13 Aug – 10 Sept	£480	£160

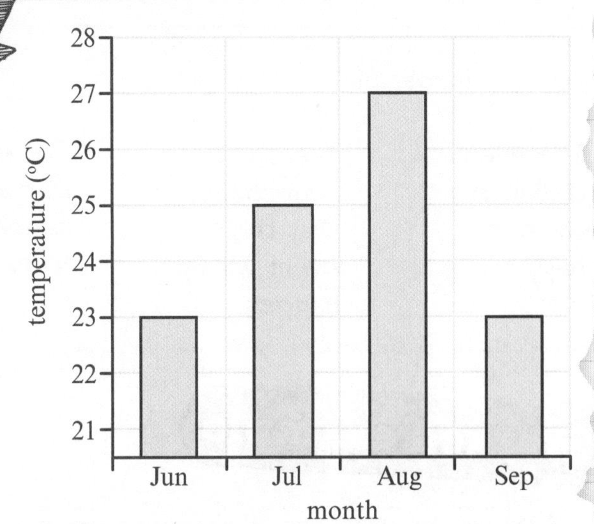

They have a budget of £1,200. When should they go?

The warmest month is August so they should try to go then. The cheapest price in August is between 13 August and 10 September. It will cost them $2 \times 480 + 160 = £1,120$ which is within their budget.

Worked example

Make a decision based on the mean. Show your working clearly.

Jason works in a shoe shop. He gets a bonus if he sells an average of £150 worth of shoes each day.

The table shows the sales he made from Monday to Friday this week.

Day	Mon	Tues	Wed	Thur	Fri
Sales	120	150	220	100	140

Does Jason get his bonus this week?

$$\text{Mean} = \frac{(120 + 150 + 220 + 100 + 140)}{5}$$

$$= \frac{730}{5} = £146$$

No, Jason doesn't get his bonus this week as the mean is less than £150

Now try this

Calculate the mean height of the plants.

Here are the heights, after 10 days, of plants grown in one type of compost:

12 cm, 15 cm, 10 cm, 8 cm, 4 cm

Elinor needs her plants to be an average of 10 cm tall. Is this compost suitable for Elinor's plants?

Had a look ☐ Nearly there ☐ Nailed it! ☐

Likelihood

Likelihood is the chance of something happening. When you leave the house in the morning you might decide whether to take an umbrella with you depending on how likely it is to rain.

Talking about likelihood

When you describe the likelihood of an event happening, use these words:

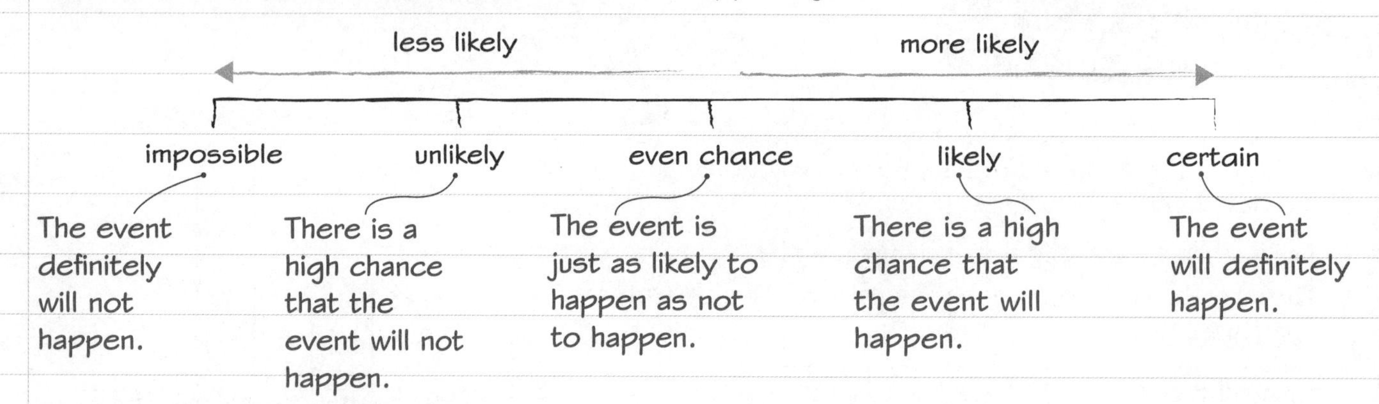

Worked example

Are these events certain, likely, even chance, unlikely or impossible?

One of your colleagues will travel to work by bus tomorrow.

It is likely because lots of people travel by bus.

When you throw an ordinary 6 sided dice it will land on an odd number.

There is an even chance because there are 3 odd numbers and 3 even numbers on a dice.

The sun will set tonight.

It is certain as the sun sets every night.

Examples of likelihood

- It is **certain** that the sun will rise tomorrow morning.
- It is **likely** that it will snow in Scotland in December.
- There is an **even chance** that a coin will land on tails if you throw it.
- It is **impossible** that an ordinary 6 sided dice will land on a 10.

Sometimes an event can happen in more than one way. A dice can land on a 1, a 3 or a 5, all of which are odd.

Now try this

Are these events certain, likely, even chance, unlikely or impossible?

- A baby chosen at random will be a boy.
- You will find a £50 note in the street tomorrow.
- Tomorrow will either be Monday, Tuesday, Wednesday, Thursday, Friday, Saturday or Sunday.
- You will see a bus today.
- The sun will fall out of the sky.

Problem-solving practice

When you are solving problems, you need to:

- ✓ read the question
- ✓ check your answers
- ✓ decide which calculation you are going to use
- ✓ make sure you have answered the question asked.

1 This table shows information about the cost of flights from Manchester to Malaga.

Date	Adult price	Child price
1 Jan–31 Mar	£140	£90
1 Apr–30 Jun	£160	£90
1 Jul–30 Sep	£240	£140
1 Oct–31 Dec	£90	£60

How much money would a family of two adults and three children save if they travel on 5 June rather than 10 August? **(4 marks)**

Tables page 66

Work out how much it will cost the family to travel at both times. Subtract the smallest cost from the largest cost to find the difference.

TOP TIP

Set out your work clearly. You can use the onscreen calculator when doing the online test, but don't forget to check each line of your working.

2 This table shows the number of hours Jasim and Mina worked during each week in February last year.

	Week 1	Week 2	Week 3	Week 4
Jasim	12	15	18	15
Mina	15	16	15	14

Compare the hours worked using the mean and the range. **(4 marks)**

Mean page 77 and Range page 78

The mean will tell you the average hours worked each week. The range will tell you how spread out the number of hours worked is.

TOP TIP

Remember to answer the question asked. Write a sentence that compares both the mean and the range.

3 The times taken for two sprinters to run 800 m are recorded below.

Sprinter A time (seconds)	Sprinter B time (seconds)
108	104
118	114
118	108
104	112

Who is the better sprinter? Give reasons for your choice. **(5 marks)**

Making a decision page 79

Calculate the mean and the range for both sprinters and decide who is faster and more consistent.

TOP TIP

The mean will tell you who has a faster time on average. The range will tell you whose times are more consistent.

Had a go ☐ Nearly there ☐ Nailed it! ☐

Problem-solving practice

A cafe owner records the types of cakes his customers buy during one day. He sells cupcakes, cream cakes and muffins.

Design and complete a data collection sheet to collect this information.

(3 marks)

Data collection sheets page 68

Draw a table with tally and frequency columns.

Type of cake	Tally	Frequency

TOP TIP

Include a tally column in your data collection sheet if you need to count how many of something there is.

This bar chart shows the number of defective parts that a manufacturing company produced in the first six months of the year.

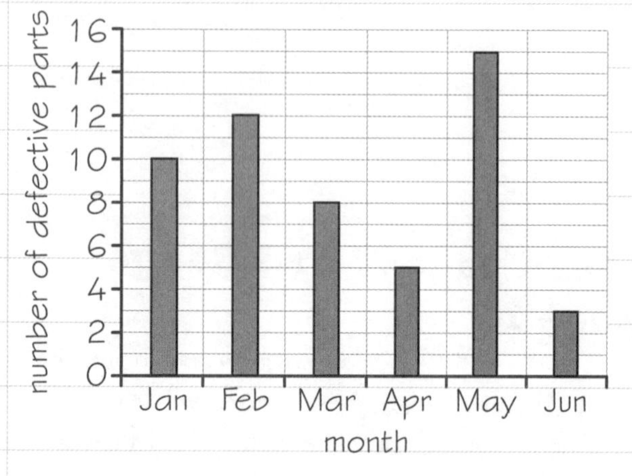

How many defective parts did they produce altogether? **(3 marks)**

Reading bar charts page 69

Read off how many defective parts were produced each month and then add to find the total.

TOP TIP

Write down each frequency before adding up to find the total.

6 These pie charts show the amount of their monthly wages Joe and Carl spend on different items.

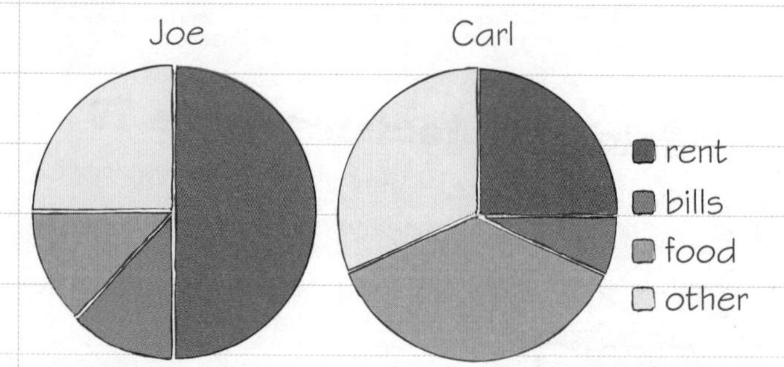

(a) Joe spends £400 per month on rent. How much does he spend on other things apart from bills and food? **(2 marks)**

(b) After looking at the pie charts, Joe thinks that he spends more money on rent each month than Carl spends. Explain why this might not be the case. **(2 marks)**

Reading pie charts page 71

Think about what information the pie charts give you.

TOP TIP

Look at the fraction of the whole pie chart each sector represents.

Problem-solving practice

This line graph shows the conversion rates between pounds and dollars or euros.

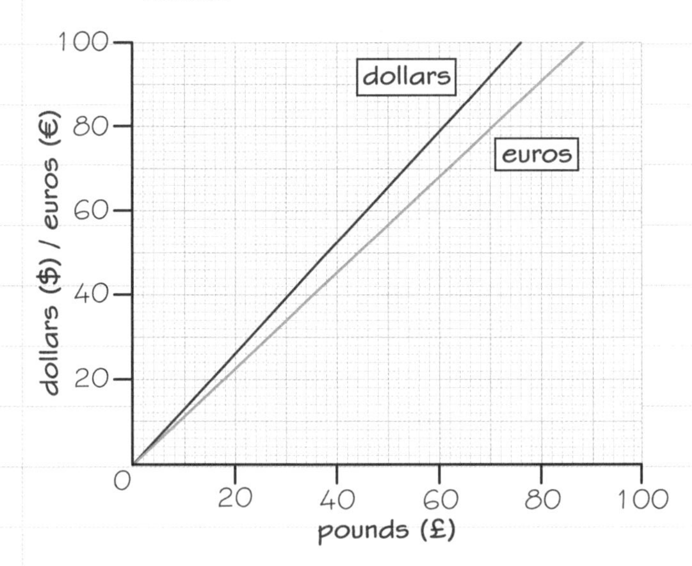

(a) How many pounds would you get for $50?

(2 marks)

(b) How many euros would you get for £70?

(2 marks)

(c) Hollie brought back $40 from a holiday in the USA. She wants to change these into euros for a day trip to France. How many euros will she get?

(2 marks)

Reading line graphs page 72

To convert between currencies, find the axis that has the currency you are converting from. Find the amount you have and read up or across until you meet the line. Then read down or across until you meet the other axis. Read off the amount.

For part (c), first change $150 to pounds, and then change the pounds into euros.

TOP TIP

Check your answer makes sense. You should expect to get fewer pounds for the same amount in dollars and euros.

This table shows information about the number of calls an online telephone support worker had over the last five months.

Year	Jun	Jul	Aug	Sep	Oct
Number of calls	40	35	60	30	15

Draw a suitable graph or chart to show this information.

(4 marks)

Planning a graph or chart page 73

Decide which type of graph or chart you will draw.

You could represent this data as a bar chart or a pictogram.

TOP TIP

If you draw a bar chart, make sure the numbers on the vertical axis are equally spaced.

If you draw a pictogram, remember to draw the key.

Answers

INTRODUCTION

1 Online test preparation

1) Click on the time icon. The time will appear in the bottom right-hand corner of the screen.
2) Zoom in or out or change the colours.

2 Online test tools

1) Because you might still get marks for correct working out even if your final answer is wrong.
2) Practise using the online tools to answer questions.

3 Using the onscreen calculator

Work out each step separately:

(a) Cost of one table:
26 + 2 = £28
Cost of three tables:
3 × 28 = £84

(b) 3 × 50 = £150
150 – 84 = £66

Use a calculator to do all the steps at once:

(a) (26 + 2) × 3 = £84

(b) (3 × 50) – 84 = £66

NUMBER

4 Number and place value

1) **(a)** three thousand four hundred and twenty-one
(b) 3 500 000
2) **(a)** 400 **(b)** 40 000
3) £30,803 £38,003 £42,023 £123,506

5 Negative numbers

1) −7, −5, −2, 6, 9, 10
2) −20°C

6 Rounding

1) **(a)** 100 **(b)** 400 **(c)** 9100
2) £39,000

7 Adding and subtracting

1) £2,443 2) £486 3) 441

8 Multiplying and dividing by 10, 100 and 1000

1) **(a)** 480 **(b)** 360 000 **(c)** 5.2
2) £7,100

9 Multiplication and division

1) 252 ÷ 7 = 36 chairs
2) 52 × 5 = 260 people
3) Profit = 40 – 18 = £22
6 × £22 = £132

10 Squares and multiples

1) 4, 16 2) 30 seconds

11 Estimating

1) **(a)** 60 + 70 = 130 **(b)** 900 – 300 = 600
(c) 50 × 10 = 500
2) 20 – 10 = 10
90 × 10 = £900

12 Checking your answer

1) 800 + 300 – 800 = 300
His answer is not close to 300 so he is incorrect.
2) **(a)** correct **(b)** incorrect (should be 362)
(c) incorrect (should be 1643) **(d)** correct

13 Word problems

Cost of entrance to zoo:
Adults: 5 × 18 = £90 Children: 10 × 12 = £120
Cost of train fare:
Adults: 5 × 16 = £80 Children: 10 × 8 = £80
Total cost: 90 + 120 + 80 + 80 = £370
So there is enough to pay for the trip.

14 Fractions

1) **(a)** $\frac{2}{5}$ **(b)** $\frac{1}{5}$ **(c)** $\frac{2}{5}$
2) two-thirds

15 Equivalent fractions

1) **(a)** $\frac{1}{7} = \frac{2}{14}$ **(b)** $\frac{1}{2} = \frac{3}{6}$ **(c)** $\frac{3}{8} = \frac{9}{24}$
2) $\frac{1}{8}$
3) $\frac{1}{5}, \frac{3}{10}, \frac{4}{10}$

16 Mixed numbers

1) $\frac{4}{3}$ 2) $\frac{50}{9}$ 3) $7\frac{1}{2}$

17 Fractions of amounts

1) **(a)** £14 **(b)** £28 **(c)** 32 g
2) $\frac{4}{9}$ of £72 = £32, $\frac{3}{4}$ of £44 = £33 so $\frac{4}{9}$ of £72 is smaller.

18 Word problems with fractions

$80 \times \frac{3}{4} = 60$ 60 × 4 = £240 20 × 2 = £40
£240 + 40 = £280 £280 – £200 = £80
David makes £80 profit.

19 Decimals

1) **(a)** 6.7 **(b)** 23.84 **(c)** 2.03
2) **(a)** 6 thousandths **(b)** 6 tenths
3) £10.30, £12.03, £12.30, £13.02

20 Decimal calculations

1) £4.39
2) 16.36 m
3) Oliver spent £56.54 and Lukas spent £56.65 so Lukas spent more.

21 Rounding decimals and estimating

1) **(a)** 4 **(b)** 16 **(c)** 10
2) **(a)** 4 + 8 = 12 **(b)** 3 + 8 – 4 = 7

22 Word problems with decimals

1) 4.45 + 2 × 1.99 = £8.43 10 – 8.43 = £1.57
Joe will receive £1.57 change.
2) Yes because 4.6 + 11.3 + 0.6 + 1.1 + 0.7 = 18.3 kg, which is less than 19 kg

23 Fractions and decimals

1) (a) 0.8 (b) 0.35 (c) 0.2 (d) 1.75
2) $\frac{3}{8}$ = 0.375 $\frac{4}{9}$ = 0.444 so $\frac{4}{9}$ is larger.
3) $\frac{5}{8}$ = 0.625 $\frac{6}{8}$ = 0.75 $\frac{4}{10}$ = 0.4 $\frac{7}{16}$ = 0.4375
   $\frac{4}{10}$, $\frac{7}{16}$, $\frac{5}{8}$, $\frac{6}{8}$

24 Percentages

1) Yes, because 100 − 32 − 7 = 61 and 61 out of 100 is 61%
2) (a) 56%
   (b) 45%

25 Calculating percentage parts

1) first: £320, second: £192, third: £96, fourth:£32
2) 330 students study German.

26 Fractions, decimals and percentages

1) 36%, $\frac{2}{5}$, 0.42
2) 5 × 12 = 60 counters in total
   35% of 60 = 21
   $\frac{2}{5}$ of 60 = 24
   60 − 21 − 24 = 15
   Hayden wins 15 counters.

27 Word problems with percentages

Gym: 45 people, 45 × 5 = £225
Swimming pool: 30 people, 30 × 6 = £180
Classes: 150 – 45 – 30 = 75, 75 × 8 = £600
Total made in the week: £225 + £180 + £600 = £1,005

28 Using formulas

dog years = 21 + 4.5 × 10
= 21 + 45
= 66 dog years

29 Ratio

1) These are not equivalent fractions as 3 : 15 simplifies to 1 : 5 not 1 : 4
2) Verity got 12 correct answers so she must have got 4 incorrect answers. This is a ratio of 12 : 4 or 3 : 1 so she is wrong.

30 Ratio problems

1) 1 : 3
2) £330

31 Proportion

(a) £2.80 (b) £6.72

32 Recipes

(a) 1200 g
(b) 16 ÷ 4 = 4
   4 × 400 g of courgettes = 1600 g
   4 × 600 g of potatoes = 2400 g
   4 × 1000 ml of stock = 4000 ml
   Simon does not have enough potatoes, so he cannot make enough soup for 16 people.

33 Word problems with ratio

(a) 40 ÷ 5 = 8 so he needs 8 tins of tuna.
(b) cost of bread rolls: 4 × 4 = £16
   cost of tuna: 4 × 3 = £12
   total cost = £16 + £12 = £28

34 Problem-solving practice

1) Wednesday £8 Thursday £3 Friday £14
   8 + 3 + 14 = £25
   It will cost more than £20, so Amina is not correct.
2) Cost of sofas
   80 × 200 = £16,000
   Sales
   Number of sofas sold: 0.75 × 80 = 60
   60 × 300 = £18,000
   Remaining sofas: 80 − 60 = 20
   20 × £250 = £5,000
   Total sales
   18000 + 5000 = £23,000
   Total amount made
   23000 − 16000 = £7,000
   Josie makes £7,000 selling the sofas.
3) coffee: 2.25 × 4 = £9
   tea: 2 × £1.75 = £3.50
   squash: 3 × 99p = £2.97
   soup: 4 × £3.60 = £14.40
   sandwiches: 5 × £3.05 = £15.25
   bread rolls: 2 × 80p = £1.60
   £9 + £3.50 + ... = £46.72
   She paid £20 + £20 + £10 = £50
   £50 – £46.72 = £3.28

35 Problem-solving practice

4) 20 ÷ 100 = 0.2, 0.2 × 1800 = 360,
   1800 + 360 = 2160
5) 70% + 20% = 90%, 100 − 90 = 10%
6) (a) 30 × 4 = 120
   (b) screws
   120 screws, 120 ÷ 10 = 12 packs of screws,
   12 × 2 = £24
   door handles
   30 × 2 = 60, 60 door handles, 60 ÷ 5 = 12 packs of door handles, 12 × 20 = £240
   total cost
   £240 + £24 = £264

36 Problem-solving practice

7) 40 + 5 × 25 = 40 + 125 = £165
   No, Johanna doesn't have enough to hire a car.
8) offer A: 4.32 ÷ 12 = £0.36
   offer B: 6.30 ÷ 18 = £0.35
   Offer B is better value for money as the cost of one packet is less
9) Claire needs: 480 g self-raising flour, 2 tsp baking powder, 160 g butter, 320 ml milk
   Claire does not have enough milk, so she does not have enough ingredients to make scones for 16 people.

TIME

37 Units of time

1) 4 × 60 = 240 minutes
2) 5 hours or 300 minutes

38 Dates

1) (a) 17 days (b) Monday 19 October
2) Thursday 4 June

Answers

39 12-hour and 24-hour clocks

1) (a) 2.45 pm (b) 14:45

2)

12-hour	24-hour
3.12 a.m.	03:12
7.50 p.m.	19:50
11.30 p.m.	23:30
10.15 a.m.	10:15

40 Timetables

(a) 06:10 (b) 48 minutes

41 Creating a time plan

One possible time plan:

10.00–11.00	Adventure rides
11.00–12.00	Arts and crafts
12.00–1.30	Wildlife show
1.30–2.00	Lunch
2.00–2.30	Bird sanctuary
2.30–3.00	Treasure hunt
3.00–4.00	Nature tour

42 Problem-solving practice

1) 1 hour = 60 minutes
$2 \times 60 + 4 \times 30 + 3 \times 15 + 1 \times 45$
$120 + 120 + 45 + 45 = 330$ minutes
$330 \div 60 = 5.5$ or 5 hours and 30 minutes

2) Yes, she will get her dress in time as there are 12 working days before the wedding.
Mon 21 Sep, Tue 22 Sep, Wed 23 Sep, Thu 24 Sep, Fri 25 Sep, Sat 26 Sep, Sun 27 Sep, Mon 28 Sep, Tue 29 Sep, Wed 30 Sep, Thu 1 Oct, Fri 2 Oct, Sat 3 Oct, Sun 4 Oct, Mon 5 Oct, Tue 6 Oct, Wed 7 Oct

3)

starts work	08:30
start of break	10:30
back at work	11:00
start of lunch	13:30
back at work	14:30
ends work	16:30

He finishes work at 16:30

43 Problem-solving practice

4) (a) 15 minutes (b) 09:45
(c) 80 minutes

5) Possible time plan

Time	Activity
10:00	data task
10:30	tour around site
11:15	inbox task
12:00	lunch
12:45	test
13:00	interview
14:00	end of day

MEASURES

44 Units

1) The units are incorrect, cm is a measure of length.
2) mm or cm
3) kg or tonnes
4) litres

45 Measuring lines

1) (a) 2.8 cm (b) 5.5 cm (c) 4.9 cm (d) 6 cm
2) Accept between 8 and 10 metres

46 Scales

(a) D
(b) B

47 Mileage charts

(a) Chartres (b) 492 kilometres

48 Routes

(a) 81 km
(b) The shortest route would be: Purple City to Pinewood Falls to Beaver Canyon to Snowdeer District to Purple City.

49 Length

1) (a) 700 cm (b) 3 m (c) 2500 m (d) 12 km
2) 60 cm, 6000 mm, 60 m

50 Weight

1) (a) 3000 mg (b) 6 kg (c) 6200 g
2) $8000 + 4000 + 30 = 12030$ g

51 Capacity

1) 5000 ml
2) 2 tins of 2 litre paint and 2 tins of 500 ml paint

52 Money

1) £4.10
2) 3 pack: £0.42 each 5 pack: £0.39 each
The 5 pack is the best value for money.

53 Temperature

(a) A is −20°C, B is −15°C, C is −25°C
(b) No, freezer B is above −18°C so is too warm.

54 Perimeter and area

1) (a) 22 cm (b) 22 cm^2
2) Any shape enclosing 20 squares

55 Area of rectangles

1) 36 cm^2
2) (a) $6 \times 4 = 24$ m^2
(b) $24 \div 4 = 6$ bags of grass seed

56 Problem-solving practice

1) 25 × 1000 = 25 000 ml
   25 000 ÷ 150 = 166.7 cups
   The number of cups that can be filled completely is 166.
2) 75 ÷ 3 = 25 g
3) Three full price cartons cost 225p.
   225 ÷ 100 = £2.25
   £2.25 – £1.80 = £0.45

57 Problem-solving practice

4) (a) 265 + 42 + 192 + 122 = 621 miles
   (b) 621 × 50 = 31050 pence = £310.50
5) (a) 41 + 36.2 = 77.2 miles
   (b)

Manchester			
43	Leeds		
97	73	Nottingham	
41	36.2	45	Sheffield

(c) Sheffield to Manchester (41 miles) Manchester to Leeds (43 miles)
Leeds to Nottingham (73 miles) Nottingham to Sheffield (45 miles)

58 Problem-solving practice

6) (a) (200 × 300) ÷ (50 × 50) = 24 so 5 packs
   (b) 5 × 6 = £30
7) 6 + 6 + 9 + 15 = 36 m   36 ÷ 3 = 12
   12 × 24.50 = £294

SHAPE AND SPACE

59 Symmetry

1) 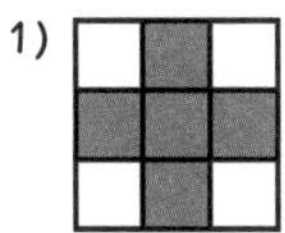

2) 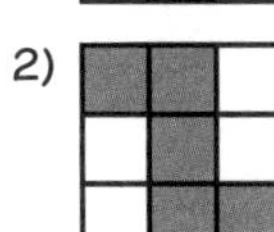

60 Properties of 2D shapes

1) square, regular   rectangle, irregular   triangle, regular
2) Yes she can.

61 Scale drawings and maps

1) (a) 250 m   (b) Yes – distance is 200 m
   (c) 8 cm

62 Using plans

1) 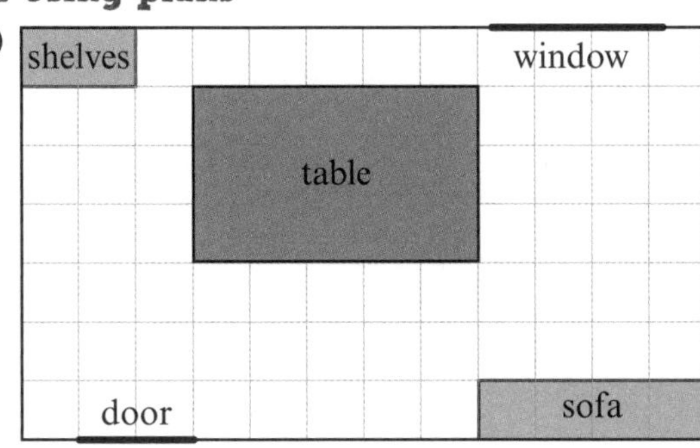

There is more than one possible answer. Check your answer meets all the requirements.

63 Angles

1) (a) obtuse angle, 136° (b) right angle, 90°
   (c) reflex angle, 300°

64 Problem-solving practice

1) 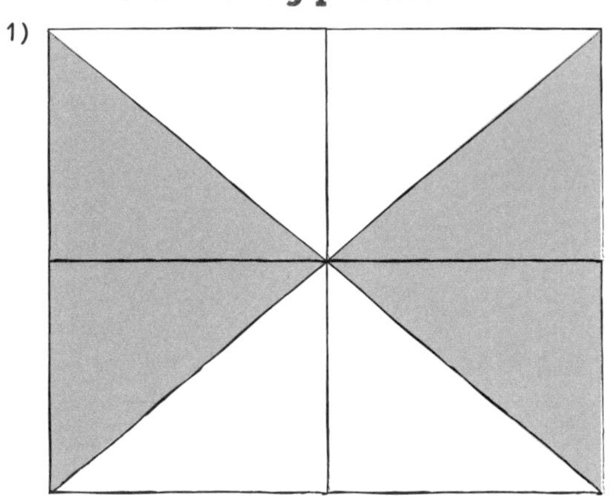

2) There are not enough paving stones because the area of the market square is 42 m × 25 m = 1050 m², which is more than 1000 m²

65 Problem-solving practice

3) (a) 160 m   (b) 80 + 120 = 200 m
4) Key:
   1 square = 0.5 m by 0.5 m
   ——— represents a wall

door
tv
shelves
window
sofa
shelves

HANDLING DATA

66 Tables

employee 1: 25 × 12 = £300
5 × 18 = £90
£300 + £90 = £390

employee 2: 24 × 10 = £240
8 × 20 = £160
£240 + £160 = £400

Employee 2 earned more money.

Answers

67 Tally charts and frequency tables

(a)

Item	Tally	Frequency
postcards	𝍸 𝍸 𝍸 𝍸 II	22
key rings	𝍸 𝍸	10
book	𝍸 𝍸 III	13
toy	𝍸 𝍸 I	11
poster	𝍸 III	8

(b) 22 + 10 + 13 + 11 + 8 = 64

68 Data collection sheets

1)

car colour	tally	frequency
red		
blue		
black		
white		
silver		
other		

2) (a)

Name	Job title	Happy	OK	Dissatisfied

(b) No – 11 people say they are dissatisfied and only 9 say they are happy.

69 Reading bar charts

(a) Any correct comment about the chart. For example: 'The type of flower that the florist sold most of was lilies.'

(b) 8 + 6 + 12 + 7 = 33

70 Reading pictograms

(a) 6

(b) 4

(c) 34

71 Reading pie charts

(a) 40 (b) You cannot tell because you do not know how many people visited Happy Hair in total.

72 Reading line graphs

£25

73 Planning a graph or chart

(a) Month will be on the x-axis and height will be on the y-axis.

(b) 0 to 26 in jumps of 2

(c) line graph

74 Drawing bar charts

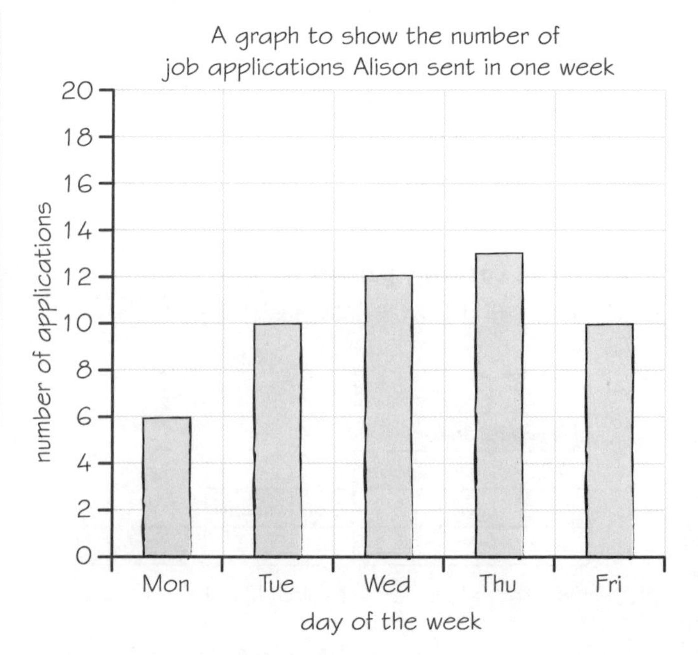

75 Drawing pictograms

1) (a)

(b) 6 goals

(c)

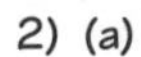

76 Drawing line graphs

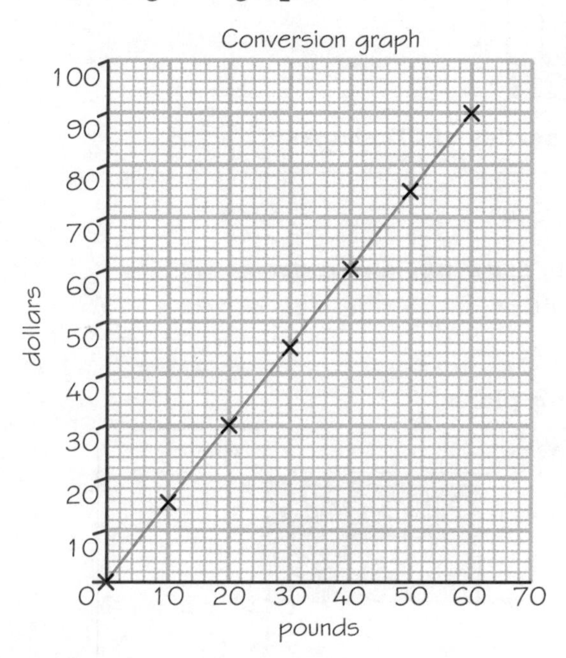

77 Mean

(a) 136 ÷ 4 = 34 (b) No, she hasn't met her target.

78 Range

80 − 20 = 60 miles

79 Making a decision

mean: (12 + 15 + 10 + 8 + 4) ÷ 5 = 49 ÷ 5
= 9.8 cm

No, the compost is not suitable as the mean is less than 10 cm.

80 Likelihood

A baby chosen at random will be a boy – Even chance

You will find a £50 note in the street tomorrow – Unlikely

Tomorrow will be either Monday, Tuesday, Wednesday, Thursday, Friday, Saturday or Sunday – Certain

You will see a bus today – Likely

The sun will fall out of the sky – Impossible

81 Problem-solving practice

1) June: 2 × 160 + 3 × 90 = £590
   August: 2 × 240 + 3 × 140 = £900
   £900 − £590 = £310
   It costs £310 less to travel in June rather than August, so the family would save £310.

2) Jasim
   Mean = 60 ÷ 4 = 15 hours
   Range = 18 – 12 = 6 hours

   Mina
   Mean = 60 ÷ 4 = 15 hours
   Range = 16 – 14 = 2 hours
   The mean is the same for Jasim and Mina. The number of hours Jasim worked is more spread out.

3) Sprinter A: mean = 112 seconds
   range = 14 seconds
   Sprinter B: mean = 109.5 seconds
   range = 10 seconds
   Sprinter B is the better sprinter. They have a lower average time and the times are less spread out.

82 Problem-solving practice

4)

Type of cake	tally	frequency
cupcake		
cream cake		
muffin		

5) 53

6) (a) £200

   (b) Pie charts don't give you information about actual amounts. These tell you the proportion of the total amount Joe and Carl earn that they spend on different things. Carl might earn more money than Joe and so the amount of rent he pays might be more.

83 Problem-solving practice

7) (a) Answers between £38 and £40
   (b) Answers between €78 and €80
   (c) €34

8) Bar chart to show the number of calls received each month

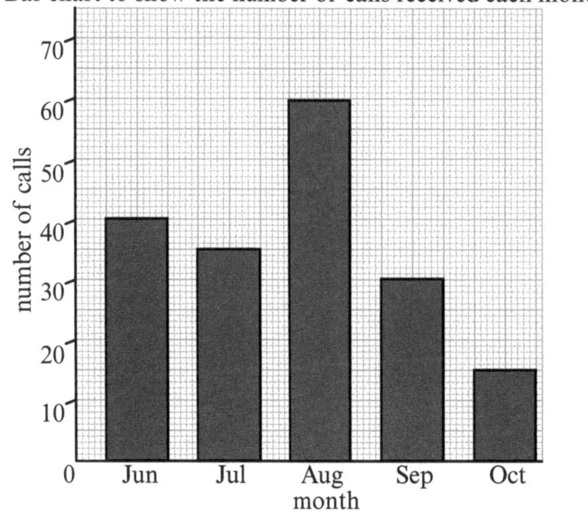

or

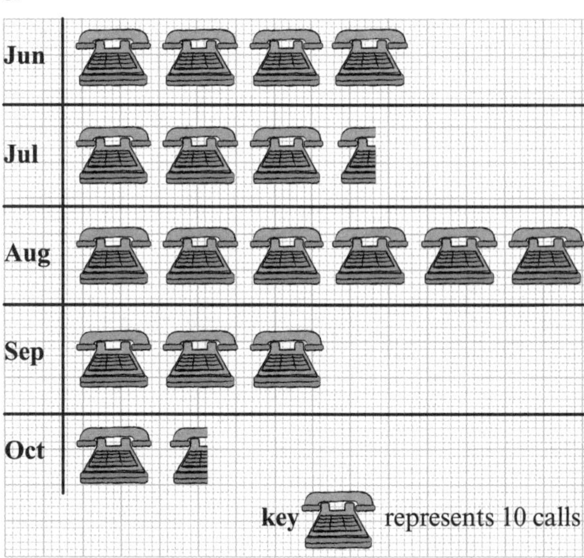

Notes

Notes

Notes

Notes

Published by Pearson Education Limited, 80 Strand, London, WC2R 0RL.

www.pearsonschoolsandfecolleges.co.uk

Copies of official specifications for all Edexcel qualifications may be found on the website: www.edexcel.com

Edited by Linnet Bruce
Typeset by Jouve India Private Limited
Produced by Elektra Media

Illustrated by Elektra Media
Cover illustration by Miriam Sturdee

First published 2016

19 18 17 16
10 9 8 7 6 5 4 3 2 1

British Library Cataloguing in Publication Data
A catalogue record for this book is available from the British Library

ISBN 978 1 292 14569 3

Printed in Italy by Lego S.p.A.

A note from the publisher
In order to ensure that this resource offers high-quality support for the associated Pearson qualification, it has been through a review process by the awarding body. This process confirms that this resource fully covers the teaching and learning content of the specification or part of a specification at which it is aimed. It also confirms that it demonstrates an appropriate balance between the development of subject skills, knowledge and understanding, in addition to preparation for assessment.

Endorsement does not cover any guidance on assessment activities or processes (e.g. practice questions or advice on how to answer assessment questions) included in the resource nor does it prescribe any particular approach to the teaching or delivery of a related course.

While the publishers have made every attempt to ensure that advice on the qualification and its assessment is accurate, the official specification and associated assessment guidance materials are the only authoritative source of information and should always be referred to for definitive guidance.

Pearson examiners have not contributed to any sections in this resource relevant to examination papers for which they have responsibility.

Examiners will not use endorsed resources as a source of material for any assessment set by Pearson.
Endorsement of a resource does not mean that the resource is required to achieve this Pearson qualification, nor does it mean that it is the only suitable material available to support the qualification, and any resource lists produced by the awarding body shall include this and other appropriate resources.